Traffic Investigation and Control

Bruce A. Hand

Golden West College

Archible W. Sherman, Jr.

Bakersfield College

Michael E. Cavanagh

University of San Francisco

CHARLES E. MERRILL PUBLISHING COMPANY
A Bell & Howell Company
Columbus, Ohio

Published by
Charles E. Merrill Publishing Company
A Bell & Howell Company
Columbus, Ohio

This book was set in Melior.
The production editor was Linda Gambaiani.

ISBN: 0-675-08610-8

Library of Congress Catalog Card Number: 75-40822

4 5 6 7 8—83 82 81 80 79 78

Printed in the United States of America

To our wives, Lee, Norma and Marie, as well as our children whose patience was unending.

Preface

Traffic control is a critically important part of law enforcement. On the national level, approximately 50,000 individuals are killed yearly in motor vehicle collisions. Over 11 million persons are disabled. Motor vehicle collisions significantly lead *all other causes of death* for individuals from ages one to forty-five. Property damage caused by motor vehicle collisions is approximately six billion dollars each year.* So the traffic officer, both in his or her role as enforcer and as investigator, is dealing in a very direct way with the preservation of human life and property.

A search for material for a traffic control class quickly revealed that there was an abundance of reference material available. However, the material was highly technical and submerged in technical and legal jargon. Most of the volumes encountered dealt with specific subjects and involved sophisticated theoretical models and algebraic equations. These volumes did not lend themselves to a general class of traffic control designed to address itself to broad areas of traffic supervision, traffic engineering, enforcement, records systems, and accident investigation.

What was lacking was a single reference which covered subjects relative to traffic control and investigation in comprehensive and understandable terms. A small number of publications were

*Statistics from *Accident Facts*, 1975 (Chicago, Ill.: The National Safety Council).

available, however, they were too simplistic and/or lacking in content.

After identifying the need for a single source of reference for traffic control and investigation, the authors felt that such a publication should have national application because of the universality of the traffic problem. Thus, the authors' efforts were directed toward sorting and assimilating a vast amount of secondary data and combining it with many years of knowledge gained through field traffic experience.

It is our hope that this book will fill the gap and provide a single reference which is comprehensive and understandable and which has national application.

ACKNOWLEDGMENTS

The authors wish to acknowledge the following for their generous assistance in the preparation of this text: The California Highway Patrol; Los Angeles Police Department; Ohio Highway Patrol; Michigan State Police; Florida Highway Patrol; Tiburon (California) Police Department; Wisconsin State Patrol; Traffic Institute, Northwestern University, Evanston, Illinois; and the School of Public Administration, University of Southern California.

Individually we would like to thank the following persons for their assistance in providing materials, suggestions, and review of the chapters, and most of all their support and encouragement in the publication of this material: Commissioner Truett Ricks of the Kentucky State Police (now on leave and presently Associate Dean of the School of Law Enforcement, Eastern Kentucky University); Sergeant Kenneth Frayer, Detroit Police Department and Schoolcraft College; former chief Maurice Lafferty and Chief Michael Manick (Tiburon Police Department); Sergeants John Braune, Steve Penkoff, and Donald Peet of the California Highway Patrol; and finally John F. Ferguson, special consultant for the Administration of Criminal Justice Systems, for his work on chapter 4, Traffic Records Systems.

Contents

1: The Traffic Problem

The success of any traffic investigation and control program is dependent upon the control of the rate and severity of traffic accidents. Critical to control of traffic accidents is the determination of their causes and remedies, which minimize their recurrence. Thus, traffic accident investigation evolves as the primary source of data for analysis of highway safety.

Historical Developments

Traffic accident investigation is nothing new. From the time a human first shouldered a burden and collided on a narrow trail with another human, traffic accidents have occurred. The taming of wild animals and their subsequent training as beasts of burden brought an increasing number of traffic accidents. The development of the wheel and the corresponding development of transportation vehicles over the ages contributed to the growth of the traffic problem.

Progress and improvement in thoroughfares paralleled advances in transportation first with trails, then roads. In ancient times, a road network consisting of hard surfaced pavements was developed by the Romans to spread their civilization. These roads were so well constructed that some of them still remain visible today.

The Romans were responsible for the development of the term "highway." During their military exploits in England, the Romans built cities on high ground above the low moors and marshes. Since they developed roads to connect their cities, they built raised roads to span the low lands. These "high ways" soon became known as highways.

Traffic accidents occurred on these roads then, even as they do now. Those investigated by the Romans led to certain control measures. Weight limits were imposed to prevent breakdown of vehicles and destruction of pavements. Restrictive laws were passed to regulate the movement of persons and property. Encroachments upon right-of-way by roadside businesses were discouraged and restricted. Roads were designed with a minimum of grades and curves to facilitate travel. In fact, it was here that the concept of the three "E's"—engineering, education, and enforcement—was discovered. Today this is still the basis for traffic investigation and control.

The methods of supervising traffic adopted during the Roman Empire were so advanced that they supplied many future generations with sufficient control for their own safety. Because there were no appreciable improvements in the means of travel, there were no real necessities to improve or alter the early practices. It was not until the twentieth century that the need arose for any major change, for it was during this period that transportation took a tremendous upswing with the advent of the automobile. This upswing was accompanied by a colossal upsurge of traffic problems which resulted in the need for new regulations and adjustments.

Rapid Growth

In September, 1899, H.H. Bliss stepped off a trolley car and was run over by a horseless carriage. It was the first time on record that a citizen had been killed in an automobile accident.[1] Since then the death on the highways has reached a proportion unequaled by war and other manmade calamities. Approximately two million persons have died in automobile accidents since the automobile's invention. Using 1900 as the base year for automobiles, in about seventy-five years the annual average death rate has been approximately 26,700 persons.

Improved roads, better signs, seat belts, shoulder harnesses, padded dashboards, collapsible steering columns, and improved tires and brakes have made driving a safer activity in the last dec-

1. Walter Pudinski, "Fatal Accident Studies," *Journal of Traffic Safety Education* 17, no. 3 (March 1971): 17.

ade. Obviously, these innovations have not eliminated traffic fatalities and driving is not as safe as it should be. Trying to scare drivers into safer actions by comparing statistics on war and highway death rates has not worked either.[2] Certainly there is a need for drivers to be careful and observe traffic laws. Possibly, this is where the emphasis should be—reminding drivers directly and bluntly of this obligation behind the wheel.[3]

Finding an Answer

There is no simplistic answer to the traffic problem. However, we are now at a point where the traffic problem has to be attacked systematically. Since, as individuals, we do not have the legal power, manpower, or resources to immediately and unilaterally eliminate the traffic problem, we must develop priority programs for corrective action collectively.[4] We must define specific problem areas, determine realistic levels of expectation, and identify those areas that can be improved upon with available resources. This can be accomplished through detailed nationwide cooperative study and analysis of traffic accidents and their causes and a systems approach to highway safety. It is through research that efforts are better directed because it supplies the necessary information for enforcement, engineering, and education programs.[5]

It is easily recognized that roads and vehicles, like other inanimate objects, can be reshaped and made over, but not so easily the driver. He or she must be governed by the rules of driving conduct in order to minimize highway chaos. Driver conduct must be guided and controlled for the welfare of all highway users. Through the years effort progressed along various lines of training, education, and by traffic laws and ordinances designed to promote orderly passage of vehicles over the streets and highways, and the relief of traffic congestion.[6]

Once attention was focused on these methods of control, the authorities and others concerned with traffic safety found themselves confronted by a variety of legal questions. By or through what agency or agencies of government could valid restrictions on use of the highways be promoted? To what extent and for what

---

2. Richard Bauman, "The Absurd Statistic," *The California Highway Patrolman* 39, no. 4 (June 1975): p. 27.
3. Ibid., p. 29.
4. Pudinski, "Fatal Accident Studies."
5. Ibid., p. 41.
6. Donald R. Drew, *Traffic Flow Theory and Control* (New York: McGraw-Hill, 1968), p. 28.

purposes could use of the streets and highways be limited or restricted? In other words, first attempts to control the mass of people through restrictive regulations affecting the driving of their own vehicles posed some highly complex and interesting questions for the courts as well as for the legislative and executive branches of government. In the area of traffic every person was concerned, either as a driver or a pedestrian, and means of control and enforcement had to be devised for a volume of offenses quite unknown before. The making of the early traffic laws followed an erratic and unrealistic course. Archaic laws and regulations, some of which aggravated the situation rather than helped it, had to be eliminated. New concepts of driver control had to be devised. New theories and practices of enforcing laws and regulations had to be contrived and developed. What had to be done was not terminal but a continuous process, because conditions change swiftly, requiring new adaptations of old principles and the development of new ones.[7]

Control of the driver posed questions not only for law-making bodies and enforcement officials, but it also called for supplementary activity on the part of educators and traffic engineers. It is our conclusion that the human element is a crucial factor in the effective control of traffic. Intelligent, efficient enforcement of well-devised regulations, with certainty and consistency, can and will offer the hope and promise of ultimate solutions to the traffic problem.

Progress toward an ultimate solution is threefold. Continuous traffic supervision effort has to be directed toward the utilization of: (1) modern enforcement methods and techniques, together with improvement in the education and training of drivers; (2) modern traffic engineering concepts which provide for the reduction of driver decision-making situations and the compatability of the highway system with driver behavior and vehicle kinematic and dynamic characteristics; and (3) realistic laws designed to prevent collision and congestion, enforced by police officers specially trained for traffic work. These laws must be administered and adjudicated by courts functioning according to modern standards for the administration of justice in traffic cases. Concentrated effort in these areas can be effective in reducing the toll of deaths, injuries, and property damage resulting from the operation of motor vehicles upon streets and highways.[8]

7. Ibid.
8. Ibid., p. 29.

Scope of the Text

Some provocative issues of traffic investigation and control have been touched upon in this introductory chapter. The purpose of this text is to look at some of these issues and at the same time explore some of the variables which have an impact on highway safety. Also, some basic concepts of traffic investigation and control, as developed through engineering, enforcement, and education efforts, which attempt to integrate people, vehicles, and highways within a relatively safe traffic environment will be explored.

The specific areas we will cover are: traffic supervision and the monumental task of identifying overall problem areas and directing resources strategically to effectively remedy the situation; traffic records systems which involve the collection of historical data of traffic problems for assessment of overall cause and effect; psychology of enforcement tactics as these relate to the psychological factors and interactions that determine whether a traffic stop is a positive, neutral, or negative learning experience; traffic directing which emphasizes the practical aspect of uniform signals and gestures, and flare patterns; traffic laws and their legitimate purpose of establishing standards of procedure to be followed by drivers in the operation of vehicles so as to promote the safe and orderly flow of traffic; traffic accident investigation and its importance in collecting and assimilating relevant information for complete analysis of traffic accident causes; and special accident investigation problems which focus on critical issues in terms of method and evidence required for the establishment of traffic violations. The last chapter gives in-depth exposure to the techniques of determining speed from skid marks.

We have attempted to expose the reader to the vast subject of traffic investigation and control. It is our hope that this text will provide the reader with an insight into the variables which have an impact on the traffic environment in terms of highway safety. Further, it is hoped that unanswered questions will provide an impetus for future research.

2: Traffic Engineering

The amount of traffic in existence in American cities at the turn of the century was amazing. As one might expect, this was not motorized traffic but animal traffic. Initially, automobiles were a rare sight and their existence was known to most people by word of mouth only. Within a few years horses, which were used extensively for transportation, began to disappear from the highways, giving way to horsepower in the form of motorized vehicles. By 1927, the transition from horses to automobiles was well underway. By the 1950s, the automobile had become recognized as an indispensable means of conveyance and transportation. Automobiles along with other types of motorized vehicles had taken full control of the highway transportation scene.

The impact of motor vehicles upon the social, economic, and political structures of modern society has been enormous. As a result, changes in social customs, business and industrial methods, and American habits and lifestyles have been dramatic. The phenomenal growth of motor vehicle use came before many were ready for it. The demand for additional streets and highways far outstripped efforts to increase highway capacity (supply). Existing highways were incapable of accommodating the increased number and use of motor vehicles. To improve the situation, vast sums of money were spent not only for the development and construction of highway systems, but also for maintenance of highways and facilities for the operation of motor vehicles which have perpetual needs.

It was not long before the interaction of three elements (highways, people, and vehicles) of the highway transportation system created a high-risk highway environment in which traffic congestions and conflicts were prevelant factors. Deficiencies inherent in each contributed to devastating traffic tolls. Traffic engineers approached this problem by straightening and grading roads, eliminating sharp turns, and establishing improved routes such as sophisticated freeways and turnpike systems. With better roads came motor vehicles equipped with bigger engines capable of greater speeds. The traffic toll did not decline. The accident patterns merely changed in type and at the same time increased in number. The pattern of cars turning over on rutty roads and sharp turns, or colliding at obscure unmarked intersections changed to colliding at high speeds with other vehicles, pedestrians, and fixed objects.

The increasing number of accidents brought a trend toward safer cars. Car tops were manufactured with more structural support. Such cars could turn over and not crush their occupants as was the case in so many cars previously manufactured without this feature. Further, padded dashes, collapsible steering wheels, anti-lock brakes, energy absorption bumpers, and safety harnesses were designed to improve car safety. As manufacturers improved cars and as these cars were driven to the extent of their speed potential, accident tolls continued upward.

Higher speeds inevitably brought a corresponding increase in traffic fatalities. Obviously, better roads and better cars did not supply the ultimate answer to the traffic problem. Authorities began to realize that whatever engineering improvements and innovations might be made or adopted, it was not the mechanical inadequacies of the motor vehicles nor the defective, outmoded highways alone which were causing the trouble. Even though these factors are an integral part of the problem, the influential effect of the driver is considerable.

Yet, the general public believes that the main reason for devastating traffic accident tolls is the deficiency of the highway systems. Traditionally, better roads through traffic engineering were viewed as being the transportation system panacea.

Through the centuries, and into the early period of the automobile, the engineering approach to highways was almost exclusively structural. The emphasis was placed on the composition of the highway and how well it supported weight rather than its geometrical features. With the advent of more sophisticated automobiles, the functional requirements of the new type of traffic gradually became recognized. Emphasis began to shift from the static aspect of road design to the dynamic factors which are so important to efficient movement. The engineering shift was toward geometric design, and during the transition a concern for aesthetic highway design developed. How well highway structures harmonized with their environment became important.

Geometric design deals with roadway elements such as cross section, gradient, curvature, sight distance, and clearances as well as with combinations of these elements. There are many traffic factors which influence geometric design. Prime consideration is given to the abilities and limitations of the driver, vehicle, and road, either individually or collectively. Also, traffic composition, volume, speed, and stopping and turning abilities of vehicles are important, especially when the demand for highway use is great and the ability of various highways to accommodate existing traffic is limited. Stated alternately, the geometry of the highway system must be in harmony with traffic performance and traffic demand in order to achieve traffic operation which is safe, efficient, and economical. To accomplish this is the goal of traffic engineering.

Traffic engineering is a relatively modern procedure for coping with the three elements of street and highway traffic: (1) the road user, (2) the vehicle and its load, and (3) the road itself. This process attempts to improve the driving efficiency of road users by reducing the number of decisions required of drivers. By improving the design and operation of roads, the traffic engineer curtails the accident potential inherent in any street or highway by avoiding the need for the driver to make a decision. Through channelization, adequate signs and pavement markings, and other devices for traffic control, the engineer removes the conflict and confusion attendant when one or more drivers are required to make a decision. The traffic engineer deals in reportable accidents (injuries and fatalities). When these accidents add up to meaningful totals they constitute a warrant for some form of traffic engineering action.

Highway traffic engineering is popularly thought of as the installation and administration of controls essential to maintaining the movement of vehicles over highway systems. Actually traffic engineering is much more. It is the combination of art and science directed toward determining traffic demand and highway capacity and determining the relationship between traffic variables, plus the application of this knowledge to planning, designing, and operating highway systems in order to achieve the safe and efficient movement of persons and goods. The essential elements of this definition are: measurement of traffic demand and highway capacity; studying and determining the relationship between traffic variables; application of knowledge to planning, designing, and operating highway systems.

Measurement of Demand and Highway Capacity

Measurement of demand and highway capacity involves measuring the traffic flow (volume). The value of measuring the flow of traffic is the determination of whether the demand is as large as the capacity during any significant length of time during a day. Volume is the number of vehicles passing a given point during a specified period of time. Volume studies are one of the processes used for traffic flow evaluation.

Volume studies are made to obtain factual data concerning the movement of vehicles and/or persons at selected points on highway systems. Volume data are expressed in relation to time such as Average Daily Traffic (ADT). The type of data collected in a specific volume study depends upon the application for which the information will be used.

Average Daily Traffic in vehicles per day is used for measuring the present demand for service on a street or highway by evaluating the present traffic flow with respect to the street system, developing the major or arterial street system, and locating areas where new facilities or improvement to existing facilities is necessary. Although ADT volume is important in determining the classification of a highway or street and designing the structural element, the direct use of it in geometric design is not practical because it does not indicate the cyclical traffic patterns.

Hourly traffic in vehicles per hour is used for determining length and magnitude of peak periods and evaluating traffic controls. Volume studies of this type are usually among the criteria for the installation of signs, signals, and markings; designation of through streets, one-way streets, unbalanced flow, and traffic routing; and prohibition of parking, stopping, and turning. Geometric design or redesign of streets and intersections is influenced by hourly traffic volume studies.

Dividing hourly volume by the average speed of traffic will determine the density of traffic. Density is the number of vehicles per mile on the traveled way at a given instance. Density may be a better measure of street service than volume since it continues to increase as congestion increases. Conversely, volume reaches a maximum under moderate congestion and then decreases with greater congestion. Density is at its maximum and volume is zero when complete blockage occurs.

Traffic studies conducted by the California Department of Transportation indicate that the greatest number of vehicles passing a given point travel at 35 mph. It was found that more traffic was expedited between the speed range of 30-35 mph than any other speed ranges. For example, on a two-lane highway with traffic using both lanes, lane capacity was most efficient (1500 vehicles per hour passing a given point) when the speed of traffic was 35 mph. On the freeway, undesirable congestion developed when traffic volume increased to the point where operating speeds were reduced below 35 mph. Speeds under 25 mph inevitably produced stopping, starting, and very jerky operation.

This phenomenon is partially attributed to the reaction of drivers in the traffic stream. The longitudinal position of vehicles in the traffic stream is the single factor which gives a driver either a sense of safety with some freedom of movement or a sense of hazard and congestion with no freedom of maneuver. At speeds greater than 35 mph, the spacing or *headway* between vehicles increases and the corresponding lag reduces the number of vehicles passing a given point. At lesser speeds, the slower pace coupled with congestion delays act to reduce the number of vehicles being measured.

Basically, congestion is the direct result of an imbalance between the supply and demand of a highway system. Congestion will occur whenever the demand exceeds capacity. In traffic engineering terms, capacity is the supply. The demand (volume) is generated by those motorists seeking to use the system. Demand can be estimated by origin and destination surveys if the times of the desired trip are obtained.

Capacity and Level of Service Criteria

The ability of a highway system to accommodate traffic is a major factor of highway capacity and level of service. Capacity is the maximum number of vehicles that can pass a point on a lane or roadway during a given period of time under prevailing roadway and traffic conditions. It is a particular rate of flow—the maximum rate. Level of service, as applied to the traffic operation on a particular roadway, refers to the quality of the driving conditions afforded the driver by a particular highway system. Variables include speed of vehicles, spacing between vehicles, and the relative interference between vehicles as well as the number of vehicles that can pass a point on a roadway in a specified period of time.

Prevailing Conditions. Roadway capacity depends primarily upon (1) composition of traffic, (2) alignment of the roadway, (3) number and width of lanes, and (4) speeds of vehicles using the roadway. The prevailing conditions are divided into two groups: (1) prevailing roadway conditions and (2) prevailing traffic conditions.[1]

Prevailing roadway conditions. These conditions are determined by the physical characteristics of the roadway and do not change unless some construction or reconstruction work is performed. Poor planning and design of a highway may result in a roadway whose prevailing conditions make it difficult to move traffic without accidents when the roadway is operated well within the limits of planned capacity.[2]

Prevailing traffic conditions. Conditions of traffic are determined upon the vehicles and pedestrians using the roadway. These conditions may change during various seasons of the year, periods of the day, and, in some locations, from hour to hour. Prevailing traffic conditions are also affected by holidays, parades, sporting events, construction, and other traffic generators or hindrances. Possibly the worst prevailing traffic conditions exist just before and immediately following major holidays.[3]

There are three levels of roadway capacity: (1) basic capacity, (2) possible capacity, and (3) design capacity.

Basic Capacity. This is the maximum number of vehicles that can pass on a given lane or roadway during one hour under the most nearly ideal roadway and traffic conditions. Actually, this is a theoretical measure, as it is most difficult to secure ideal conditions when roadways are subject to maximum use.[4]

Possible Capacity. This is a measure of the maximum number of vehicles that can pass a given point on a lane or roadway during one hour under the prevailing roadway and traffic conditions.

---

1. J. Stannard Baker and William R. Stebbins, Jr., *Dictionary of Highway Traffic* (Evanston, Ill.: Traffic Institute, Northwestern University, 1964).
2. Paul B. Weston, *The Police Traffic Control Function*, 2d ed. (Springfield, Ill.: Charles C. Thomas, 1964).
3. National Commitee On Uniform Traffic Laws and Ordinances, *Uniform Vehicle Code and Model Traffic Ordinances* (Washington D.C., 1972).
4. Committee on Highway Capacity of the Highway Research Board, *Highway Capacity Manual* (Washington, D.C.: U.S. Bureau of Public Roads, Government Printing Office, 1960).

This is a positive measure or quantity. It cites a traffic volume that cannot be exceeded without changing one or more of the prevailing conditions.[5]

Design Capacity. This is the number of vehicles that can pass over a given section of a roadway in one direction during one hour under specified traffic conditions and operating at a "level of service." A design capacity is a volume generally selected for design purposes which will provide a desirable level of service. This definition replaces the concept of "practical capacity." The "practical capacity" concept dealt with the maximum number of vehicles that could pass a given point per hour without traffic density being so great as to unduly restrict the driver's freedom of maneuver under prevailing roadway and traffic conditions, and the maximum number of vehicles that could enter and clear an intersection without waiting for more than one complete signal cycle.[6]

These terms for roadway capacity levels have been found to be more applicable for rural conditions where higher speed is desirable. The desire, in urban conditions, for a more dependable and higher quality of service has brought about a change in the application of "capacity" in the design and operation of streets and highways. This new application has been termed the "level of service" concept.

Level of Service. Level of service is a qualitative measure of the effects of a number of factors which include speed and travel time, traffic interruption, freedom of maneuver, safety, driving comfort and convenience, and vehicular operational costs. Each of these factors is somewhat related to all the others. The volume of traffic using a facility affects all the factors and, in general, the greater the volume, the more adverse are the effects. As the ratio of the volume of traffic on a highway and the ability of the highway to accommodate traffic approaches unity, congestion increases.[7]

Both capacity and level of service are functions of the physical features of the highway system and the interaction of vehicles in the traffic stream. The distinction between them is that a given lane or roadway may provide a wide range of levels of service but only one possible capacity. The various levels for any specific roadway are a function of the volume and composition of traffic. A given lane or roadway designed for a given level of service at a specified volume will operate at many different levels of service

5. Ibid.
6. Ibid.
7. Ibid.

as the flow varies during different periods of time. In other words, fluctuations in demand do not cause fluctuations in capacity but do effect changes in the quality of operation afforded the driver. In a very general way, then, highway planning, design, and operational problems become a question of whether a certain roadway capacity can handle the projected or measured demand at an acceptable level of service.

There are various factors which inhibit the ability of a highway system to accommodate traffic. The most important factors that may reduce highway capacity are:

1. Lanes less than twelve feet wide have a lower capacity.
2. Retaining walls, bridge abutments, parked cars, and the like all reduce the effective width of a traffic lane, cause lane-stradling and result in lower capacity.
3. Lack of adequate shoulders also reduces the effective width of a traffic lane by causing vehicles to shy away from the edge of the pavement and travel toward the center of the roadway. Greatest reduction in capacity is when the width of the shoulders of a road will not permit the parking of a disabled vehicle out of the traffic lane.
4. Commercial vehicles occupy a greater road space and influence other traffic more than other passenger cars. They are slow moving, especially on grades, and are slower to close gaps.
5. When drivers are restricted in their freedom of movement by restricted sight distances the capacity of the road will be lowered.
6. Controlled intersections impair highway capacity by depriving the traffic stream of a portion of the time during which it would otherwise be on the move.

Studying and Determining the Relationship Between Traffic Variables

Gap Acceptance

One of the most important aspects of traffic operation is the interaction of vehicles within a single stream of traffic or the interaction of two separate traffic streams. This interaction takes place when a driver changes lanes, merges into a traffic stream, or crosses a traffic stream. Inherent in the traffic interaction associated with these basic maneuvers is the concept of gap acceptance.

Gap acceptance is best illustrated in the analysis of freeway ramp connections (fig. 2–1).

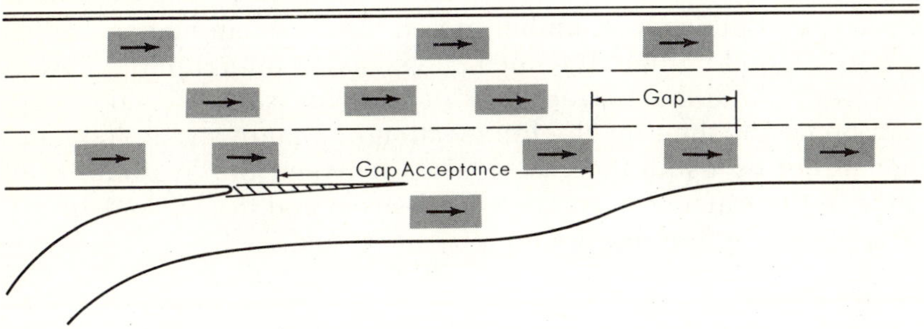

FIGURE 2–1
Gap Acceptance

The unsignalized intersection, another location in which gap acceptance provides the basis of operation, is essentially just a special case of ramp-freeway connection in which there is no acceleration lane.

Unoccupied road spaces ahead of each vehicle determine the distribution of vehicles in any one-way traffic stream. The road spaces are commonly referred to as gaps. A gap is the elapsed time between the arrival of successive vehicles on a roadway at a specified reference point (see fig. 2–1). Gaps are measured from head to head of successive vehicles. The size of a gap is usually stated in feet (when stated in units of time it is termed a "headway"). Each gap includes its own vehicle length; thus, one-half mile of roadway includes gaps totalling one-half mile in length. That portion of the current gap which remains when a vehicle on a minor street arrives at an intersection is called a lag.

Variations in gap length in a traffic stream may range from a few feet in bumper-to-bumper traffic to relatively great distances during the early morning time period. Gap length is related to the overall density at the time of traffic stream observation. At lower speeds gaps have a tendency to be fairly uniform, but as the speed of the traffic stream increases, so does the variation in gap length.

Gaps of adequate lengths provide a driver with an opportunity to either accept them or reject them. A gap is accepted when the driver crosses or enters the traffic flow between two vehicles on the roadway. Gap acceptance is based on the probability that a gap is acceptable to a vehicle waiting to enter or cross the traffic stream (both acceptable to the driver and present in the traffic stream).

It is possible to determine gap acceptance for a specified intersection and hence the number of vehicles that can be handled on a minor street without the need of a signal to interrupt the flow of the traffic on a main street. Justification for traffic signal control can be established for the intersection of a main street and a minor street by using the relationship between the availability of gaps in the traffic stream on a main street and the lag and/or gap acceptance by the drivers on a minor street.

Speed Zoning

Speed zoning is a traffic engineering tool used to derive the best traffic service for a given set of conditions. Generally, speed zoning is applied to major highways carrying appreciable volumes of traffic; transition points on major highways from rural to urban conditions; areas of high accident frequency attributable to excessive speed; and areas with unusual enforcement problems. Low volume streets offer little opportunity for encouraging closely grouped speeds due to the absence of traffic platoons under normal conditions. The basic intent of speed zoning is to influence as many drivers as possible to operate at or near the same speed—thus reducing conflicts created by wide differentials in operating speeds.

Speed zoning is used to establish speed limits which adher to the principle of the Basic Speed Law. The Basic Speed Law is the basis for speed regulations in all fifty states.

> No person shall drive a vehicle upon a highway at a speed greater than is reasonable or prudent having due regard for weather, visibility, the traffic on, and the surface and width of, the highway, and in no event at a speed which endangers the safety of persons or property. [8]

Inherent in the Basic Speed Law is the driver's recognition of driving conditions, which vary widely from time to time and place to place. No set of fixed driving rules can adequately serve all conditions. The Basic Speed Law is founded on the belief that

8. State of California, *Vehicle Code* (Sacramento: Department of Motor Vehicles, 1973).

most motorists are able to modify their driving behavior properly, as long as they are aware of the conditions around them.

Many times, unfortunately, drivers are unaware of all the conditions around them. Because of this, engineering and traffic surveys are needed to establish certain speed zones. An engineering and traffic survey is a survey of highway and traffic conditions which includes but is not limited to: (1) prevailing speeds as determined by traffic engineering measurements; and (2) accident records; highway, traffic, and roadside conditions not readily apparent to the driver.

The data collection phase of speed surveys is extremely important and requires considerable care due to the many variables involved and the source of possible bias in sampling (sample size should consist of at least 100 select observations). Some common errors that tend to introduce bias, and the procedure for eliminating these errors, are:

1. Selecting the first vehicle in a platoon of traffic. When traffic is constantly platooned, vehicles from varying positions in the platoon should be selected. In platoons which are densely packed, the traffic may be too heavy to permit a good survey and should be avoided.

2. Selecting too large a proportion of trucks. The same proportion of trucks in the sample as exist in the traffic stream should be obtained.

3. Selecting too large a proportion of higher speed vehicles. Ignoring measurement of normal speed vehicles to measure high speed vehicles will result in the survey being biased toward the upper speed ranges. Normal speed vehicles should be considered.

Important to speed surveys are inventories of road conditions and accident records. Information about existing speed zones, prevailing speeds, location of accidents, roadway alignment, traffic volumes, and other general roadside physical characteristics should be obtained. Visibility restrictions should be recorded and deficiencies corrected to assure compatibility with the prevailing speed.

When a concentration of reported accidents or an accident rate significantly higher than normal for the type of roadway under study is discovered, a detailed accident analysis should be made, including a collision diagram for the route or intersection as necessary. Adequate consideration may then be given to other corrective measures, the degree of enforcement needed, and the general applicability of any posted speed limit at all.

Analysis of Speed Survey Data

Two characteristics developed from prevailing speed data are of primary importance in the selection of a reasonable speed limit: the critical (85th percentile) speed and the pace. The critical speed is the speed at or below which 85 percent of the traffic is moving. The critical speed can be determined directly from a field speed data sheet specifically designed for speed surveys by counting from the top speed the number of vehicles equaling 15 percent of the total number of vehicles observed. Experience has shown that the 85th percentile speed is the one characteristic of high speeds which most nearly conforms to a safe and reasonable limit.

The 85th percentile speed is usually within two miles per hour of the upper limit of the pace speed. The pace is the 10 mph range of speeds containing the largest number of observations. A normal speed distribution will contain approximately 70 percent of the sample within the pace with 15 percent above and 15 percent below. These are plotted on a speed zone survey sheet (see figures 2–2, 2–3, and 2–4), vehicle speed data sheet, and a cumulative speed curve graph.

As a final aid to establishing realistic speed zones, the following practical considerations should be kept in mind:

1. Intermediate speed limits are applicable to through routes having positive intersection controls, good signing, stripping, and markings to accommodate appreciable volumes of traffic from beyond the immediate neighborhood.
2. Unusually short zones of less than a half mile in length should be avoided whenever possible.
3. Speed zone changes should be coordinated with visible changes in roadway conditions or roadside development.
4. Successive 5 mph speed zone changes should be avoided by properly selecting longer speed zones in 10 mph increments.
5. Speed zoning should be coordinated with adjacent jurisdictions to assure compatibility.

An after-study of operating speeds in a newly established speed

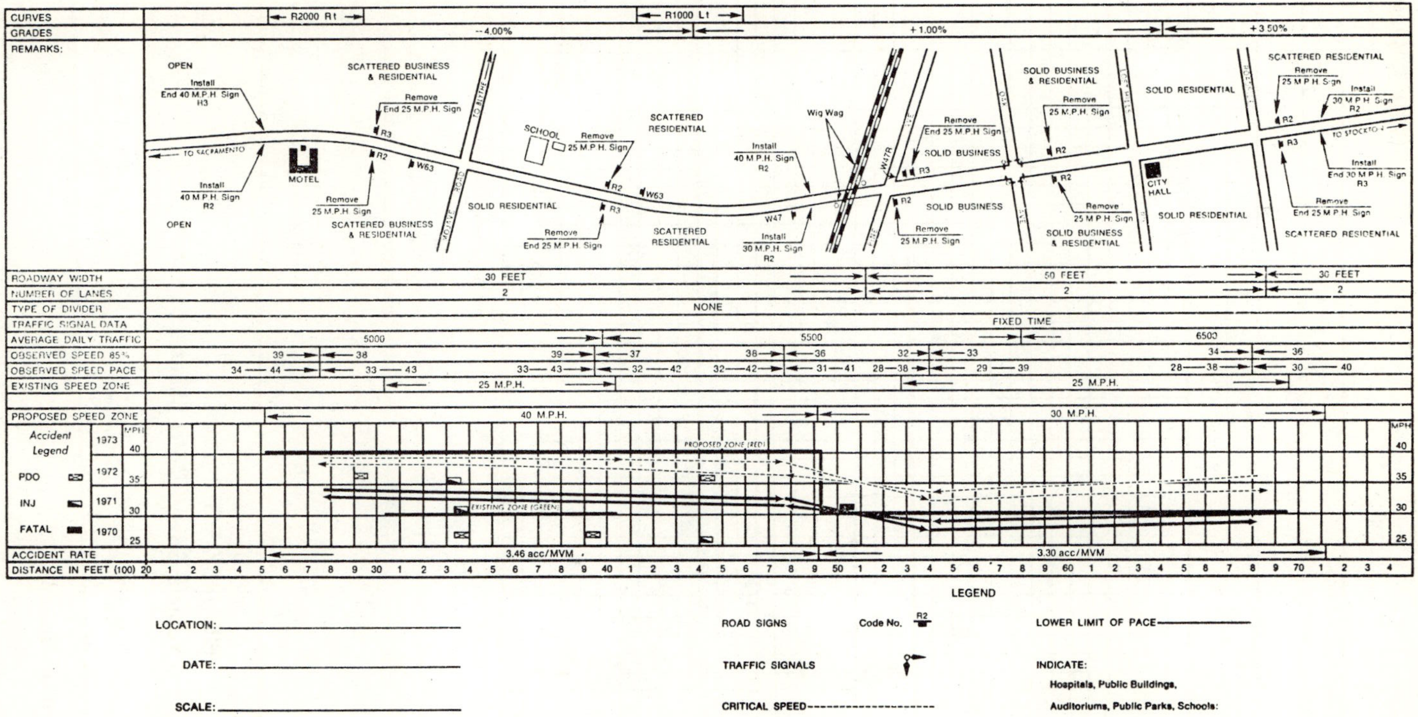

FIGURE 2–2
Speed Zone Survey Sheet

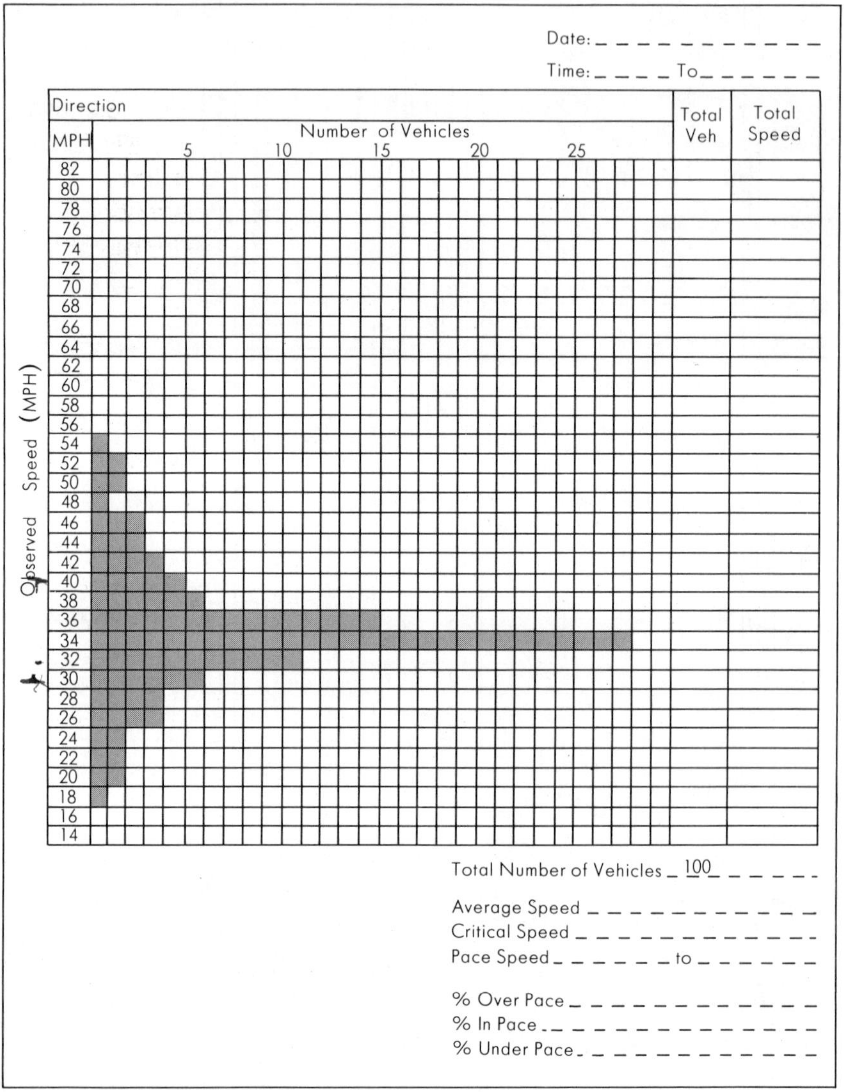

FIGURE 2–3
Vehicle Speed Data Sheet

zone should be made to verify appropriateness, relative effectiveness, and general acceptance by the motoring public.

Types of Roadways

Local Street. The function of local streets is to provide access from a collecting street to abutting land and properties. The characteristics of local streets are (1) low speeds, (2) low volume, (3) access to abutting lands, (4) space for turning moves to give access to local establishments, and (5) curb parking.

Collector Street. A street which serves the internal traffic movement within an area of the city, such as a subdivision, and connects with the major arterial system. A collector street does not handle long through trips and is not continuous for any great length.[9] While speeds are as low as local streets, a collector street handles a greater volume of traffic, though providing only limited curb parking.

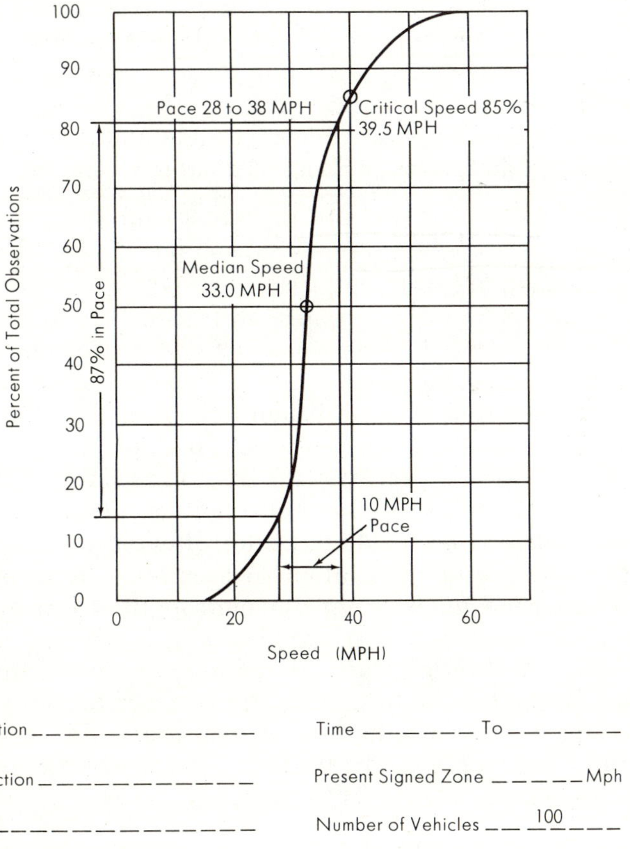

FIGURE 2–4
Cumulative Speed Curve

9. Baker and Stebbins, *Dictionary of Highway Traffic.*

Major Arterial Highway. This street serves those major movements of traffic within or through the metropolitan area not served by expressways. Major arterials interconnect the principal traffic generators within the city and the important rural routes.[10] While streets in this category may serve abutting property, the primary purpose of these roadways is to provide uninterrupted movement of a high volume of through traffic. They are characterized by access control and no parking.

Expressway. An expressway is a divided arterial highway for through traffic with full or partial control of access and generally with grade separations at intersections.[11] Characterized by the uninterrupted flow of large volumes of traffic at high speed over long distances between major traffic generating areas or through a city or other area of congestion, these roadways not only have limited access and no parking but also have adequate deceleration and acceleration lanes as well as adequate shoulders or turnouts for the emergency parking of disabled vehicles.

Freeway. Freeways provide for movement of large volumes of traffic at high speeds safely. In most cities, the surface-street system, composed of locals, collectors, and major arterials, is proving inadequate in handling expanding traffic volumes. The streets create congestion at intersections, imperil pedestrians, and create conflicts among vehicles moving and parking. Freeways represent a major innovation in road design to cope with these conditions.

Freeways have emerged as facilities specifically created to move large volumes of traffic at high speeds safely. They perform two main functions in the metropolitan area. They separate through traffic from local traffic to relieve congestion, and they provide rapid and convenient accessibility between different parts of the area and between one metropolitan area and another.

A freeway is a divided arterial highway for through traffic. This preference to through traffic is achieved by providing specially designed ramps that allow traffic to enter or leave at designated points and by prohibiting crossings at grade through the use of over- and underpasses. The biggest departure of the freeway from the surface street design is in its controlled access feature.

Traffic Lane. A strip of roadway intended to accommodate a single line of forward-moving vehicles is a traffic lane. Lanes may be as narrow as nine to ten feet on low-speed streets, but eleven

---

10. Ibid.
11. Ibid.

or twelve feet—or more—is required on highways set aside for higher traffic speeds.

The distribution of vehicles into lanes is vital to the capacity of any given roadway. The basic factor in this distribution is the geometric design of the roadway, but the condition of the road, the width and number of lanes, parking, accidents, and disabled cars all influence traffic streams and their distribution into lanes.

Traffic Stream. The name for the traffic, usually vehicular, moving in one or more lines in the same direction.[12] The operators of motor vehicles act in such a manner that it is difficult to average their reactions while part of the traffic stream. Each driver is a separate and distinct unit, having only limited coherence to other units on the street or highway.[13]

The longitudinal position of vehicles in the traffic stream is the single factor which gives a driver a sense of safety with some freedom of movement or a sense of hazard and congestion with no freedom to maneuver. As the traffic volume increases each driver is affected more and more by his fellow motorists, with the high-speed driver feeling this restriction more than the driver operating at normal speeds.

Conflict Patterns

Lane Changing. An operator who moves from one lane to another because of diminishing gap length has made a forced lane change. Any other change of lanes is termed an optional lane change. These terms give a general indication of the actions of the driver. When a driver is faced with a rapidly closing gap between his or her vehicle and the car ahead, he or she finds the longitudinal position of his or her vehicle unsatisfactory. He or she may accept gaps in adjacent lanes as short as 125 feet at speeds from 30 to 40 mph. Drivers making optional lane changes look for longer gaps before they move from their lane of operation.

The maneuver of overtaking and passing the car ahead when approaching zero headway also requires lane changing. The first movement places the overtaking vehicle in the adjacent traffic stream and the second movement returns the passing vehicle to its original lane of travel.

In either case a definite conflict occurs. This conflict in lane changing usually results from lack of opportunity to move into the adjacent lane. When the change is made in lanes of a traffic

12. Ibid.
13. Weston, *Police Traffic Control Function.*

stream moving in the same direction, the conflict is mainly with the rear vehicle in the adjacent lane. When the lane change is made as a part of an overtaking and passing maneuver there is a greater potential conflict in the oncoming vehicle. A lesser conflict exists at all times in lane changing between the overtaken vehicle and the overtaking or lane-changing vehicle.

Possibly the greatest enemy of effective lane distribution and the most effective creators of conflict are lane-straddling drivers. These individuals test the patience of fellow drivers in the traffic stream and offer an excellent opportunity for police remedial action. Normally lane-straddlers will make up about 20 percent of the volume on a highway. While a certain amount of this percentage will be operators efficiently shifting from lane to lane for legitimate reasons, the great majority are true lane-straddlers.

The careless driver straddling two lanes of a one-way traffic stream is an offender literally screaming for police attention, as he or she often causes two lanes of traffic to line up behind his or her vehicle. Next in importance in relation to the traffic stream, but possibly most important in minor accident causation, is the driver who timidly waits in his or her lane of travel until an oversize gap appears in the adjacent lane of travel. Vehicles also line up behind this driver and create increased conflict because of impatient drivers in cars behind, who dart into the adjacent lane, accepting unusually short gaps. A driver's acceptance of a gap so short that he or she not only influences the car behind but also forces that car's operator to brake creates a sudden deceleration hazard in the adjacent lane. Such action often results in rear-end collisions of a pile-up nature, with the offending operator blithely moving away, unless he or she has been in contact with the suddenly decelerating vehicle.[14]

Merging Traffic. This is the process by which drivers in two separate traffic streams moving in the same general direction combine or unite to form a single stream.[15] Merging traffic cannot be accomplished until an adequate gap occurs in the lane into which the merging driver wishes to move. An ideal merging maneuver should be carried out without retarding the rear vehicle concerned. While most delays occur to the merging vehicles, there is also delay to vehicles in the lanes of traffic into which the merging takes place.

---

14. Ibid.
15. Ibid.

When the merging vehicle is stationary (as occurs when entering the traffic stream after a required full stop) or moving slowly, average drivers require larger gaps for merging; but when the merging vehicle's speed approaches that of the rear vehicle, relatively short gaps are accepted for merging by most drivers. Acceleration and deceleration lanes on modern highways provide space for merging vehicles to adjust speeds for the purpose of entering or leaving a fast-moving traffic stream.

All merging maneuvering has a definite conflict area. This area begins at a safe distance back of the collision area and extends to a point beyond where the merging vehicle has gained approximately normal speed. The collision area extends from the point of entry of the merging vehicle to the far end of the area of conflict.

Diverging Traffic. The diverging vehicle develops an area of conflict in the lane from which it departs, and when this maneuver is made from the wrong lane of a multilane roadway the inherent hazard of leaving a highway is increased and the conflict area extends from the point of initial movement to the point where the diverging vehicle clears the path of the rear vehicle.

Even when made from the correct lane this diverging maneuver creates some conflict unless traffic volume is very low, as the diverging vehicle must slow up in order to make a turn. This creates the possibility of a rear-end collision as long as the turning vehicle remains in its original lane of travel.

The Crossing Maneuver. No greater conflict exists in the time-space relationship of vehicles to each other than in this maneuver. The conflict area extends some distance from the intersection or crossing. The speeds of the vehicles concerned, the type of vehicle crossing contemplated (right-angle, oblique, and opposed left-turn), and the volume of vehicles in each flow determine relative degrees of interference to crossing traffic. Safety at the intersection may require some form of traffic control. Control lessens the capacity of the entering traffic streams but reduces conflict in the interests of safety.

Intersections. Intersections are heavy conflict areas in which collisions may occur; for this reason they are of particular interest to persons engaged in accident reduction. The normal right-angle intersection of two-way streams of traffic develops thirty-two conflict points, of which half are of the collision potential crossing type (see fig. 2–5). The number of potential conflicts per hour is dependent upon vehicle volume in each flow of traffic. When the entering stream carries 200 vehicles per hour, approximately 1200 potential conflicts are created in each hour.

Traffic engineering techniques are available to minimize traffic conflicts. For example, the construction of left-turn lanes controlled by traffic signals can reduce conflicts at intersections and improve pedestrian safety by providing turning priority for drivers and favored crosswalks at each intersection entrance for pedestrians.

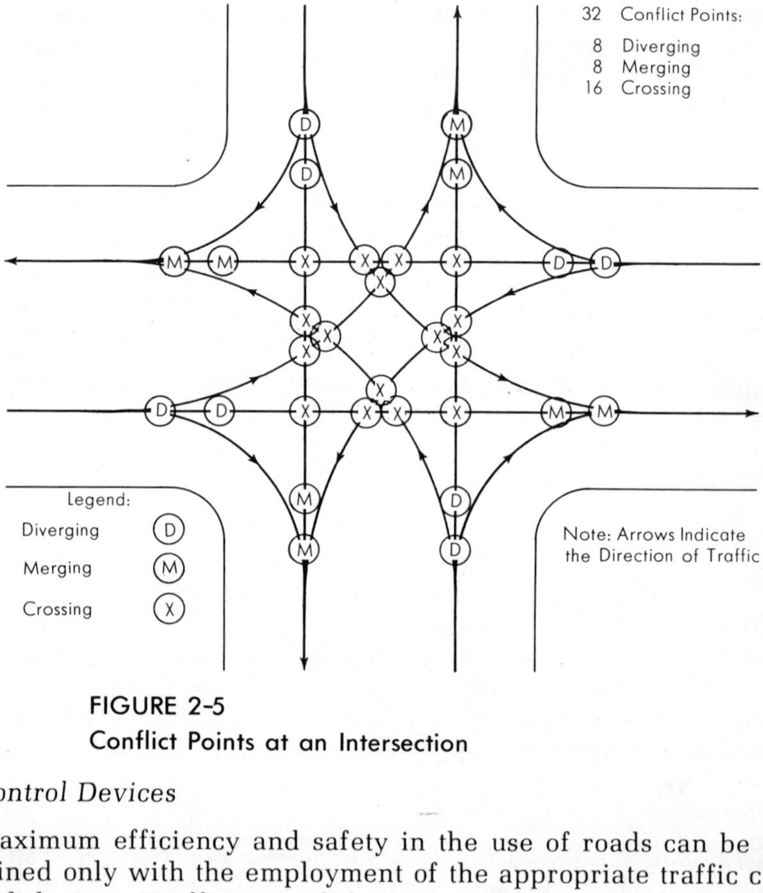

FIGURE 2-5
Conflict Points at an Intersection

Control Devices

Maximum efficiency and safety in the use of roads can be obtained only with the employment of the appropriate traffic control devices. Traffic control devices are all pavement markings, traffic signals, signs, and markers placed on, above, or adjacent to a roadway by authority of officials having the responsibility to regulate, warn, or guide traffic.

Traffic control devices are basically safety devices. Their function is to guide the motorist to his or her destination, warn of potential hazards, point the right of way, channel traffic, and otherwise regulate conduct in the traffic stream.

There are three types of traffic control:

1. Regulatory controls have the authority of law and impose precise requirements upon the road user.

2. Warning controls inform road users of potentially hazardous road conditions or unusual traffic conditions, not readily apparent to them, and thus impose responsibility upon the individual driver to employ caution as he or she proceeds through the danger area.

3. Guiding controls inform road users of route designation and other pertinent information.

These three types of traffic control devices point out the overall objective in their use—movement with safety. The individual driver is regulated in accordance with existing rules and regulations, warned of hazards and expected to comply with such warnings so that he or she does not create accident-prone situations while searching for the best route to his or her destination.

There are six elementary requirements for any traffic control device: (1) it should be capable of fulfilling an important need; (2) it should command attention; (3) it should convey a clear, simple meaning at a glance; (4) it should command the respect of road users; (5) it should be located to give adequate time for response; and (6) it should be sanctioned by law.

All traffic control devices are basically visual communication techniques. The traffic engineer uses signing as the principal means of visual communication with the driver of a vehicle. This signing must be designed to provide adequate visibility distance to enable the driver to respond and take positive action. To avoid hazardous situations, careful consideration must be given to sign design and to sign placement with respect to the roadway. Design of the signing must assure that such features as size, contrast, colors, shades, composition, and lighting or reflectorization where needed are combined to draw attention to the device; that shape, size, colors, and simplicity of message combine to produce a clear meaning; that legibility and size combine with placement to permit adequate time for response; and that uniformity, reasonableness, size, and legibility elicit compliance.

In order for the driving public to continue to respect these control devices, it is important that they be promptly removed from the public highway when no longer required or if for any reason they tend to encourage disrespect for control devices in general.

Traffic Engineering Application

Accident reduction can be achieved by correcting engineering deficiencies which contribute to accidents. The key is the remedial treatment of locations where the roadway environment is contributing to an accident hazard. The engineer must relate the physical and environmental characteristics of each location to the accident pattern. He or she must have firsthand knowledge of the signal head locations, street width, sight restrictions, lane marking (or lack thereof), grades, crown, and particularly any physical characteristics that make this location different from similar locations in the community. He or she should also be aware of traffic volumes and turning movements, parking restrictions, signal timing, composition of traffic, lining up tendencies, and other environmental characteristics which may have an effect on accident hazards.

The most effective way of gathering this information is to analyze accident records while observing traffic operations at the study location. The role of enforcement officers in such analysis cannot be overemphasized. Frequently, local police officers who have attended accidents at a given location can provide insights about the factors which may have caused accidents. Although each location will present unique hazards, a partial list of factors follows which will enable recognition of some common techniques for reducing hazards.

Primary Collision Factors	*Corrective Measures*
Left-turning vehicles struck by opposing cars (not at an intersection).	This pattern is usually found in areas where there is a lack of driveway control. Clusters of these accidents usually are found at drive-in restaurants, bowling alleys, or other traffic generators of this type. Such accidents can be reduced by installation of curb and gutter to prohibit high speed, low angle turning movements.
Right-turn accidents.	If a heavy volume of right turns is found in association with a

Primary Collision Factors	*Corrective Measures*
	high number of right-turning accidents, provision of right-turn lanes and channelization can frequently reduce this hazard.
Rear-end collisions involving left-turning vehicles (at an intersection).	These accidents can be reduced by provision of left-turn storage lanes. Such lanes can frequently be created by removing parking at an intersection or by shoulder widening where sufficient width cannot otherwise be found to create such turning refuges.
Left-turning vehicles struck by on-coming cars (signalized intersections).	If sufficient left-turning volumes are found, this type of accident can be reduced with a separate signal phase to protect left-turning vehicles.
Right-angle collisions (nonsignalized intersections).	Right-angle collisions are usually associated with substandard sight distance or lack of acceptable gaps in through street traffic. There are three common solutions to these problems: 1. If a signal warrant is met, installation of a semi-actuated traffic signal will reduce this type of collision. 2. Removal of obstructions which restrict the view of oncoming traffic. This treatment will permit the driver to choose a safe crossing gap. 3. Channelization may be used to reduce the conflict area of the intersection, thereby increasing the number of acceptable gaps available to crossing vehicles.
Rear-end collisons.	A pattern of rear-end collisions, usually on a downgrade, at

Primary Collision Factors	*Corrective Measures*
	locations approaching stop signs or traffic signals can often be attributed to a low coefficient of friction. A rough-textured seal coat can be used to reduce this hazard.
Right-angle and rear-end collisions (signalized intersections).	A combined pattern of this type usually can be attributed to poor signal timing or poor signal head visibility. Utilization of dual, far side signal indications can reduce this hazard. If approach speeds are relatively high or if long lines of traffic can be expected, use of 12-inch signal lenses can also effect an improvement.

The most important engineering facts to be gained from accident data are physical location and type of collision, both of which can be important indicators of deficiencies of the roadway system. Traffic engineers use accident reports to study physical locations which have been identified as hazardous to see if a pattern of accidents exists. If such a pattern exists, it is compared to the physical elements at that location to see if some element is causing the pattern.

Engineers actually need very little information from each accident report. But they do need accurate and complete reporting on the direction of travel, time of day, day of week, unusual pavement conditions, noninvolved vehicles, and driver intent. The need for reporting noninvolved vehicles and driver intent will be discussed in greater depth as they are the two elements frequently not mentioned on accident report forms.

Identifying Hazardous Locations

Traffic accident data must be filed by location for use by engineers. Accident reports should be filed either by intersection, or other location identities, such as street number in front of where the collision occurred, or highway mile post markers.

Selecting Locations for Study

One method is to select for analysis all intersections having ten or more accidents in a twelve-month period. If the number of intersections selected by this method is less than fifteen or more than forty, some other number of accidents per location should be used as a guide.

An alternative method assigns a rating number to each accident, and the weighted result is used instead of total accidents. Each fatal accident is given a factor such as 12, each personal injury accident rates 3, and each property damage 1. This method recognizes accident severity as a prime factor in selecting intersections to be studied.

A combination method is also possible. An intersection would qualify for study if, for example, it showed (a) ten or more total accidents or (b) a weighted factor, based on severity, of twenty or more. In this way, a location with twelve property damage accidents and one personal injury accident would qualify for study under (a), while another with one fatal, two personal injuries and three property damage accidents (total 21 points) would qualify under (b). Either method separately, or the combination of both, may be used as desired.

It is not uncommon to find intersections in medium-sized cities with thirty to forty accidents per year. Such experience indicates needed improvement, and the accident reports will reveal some of the probable changes required. The accident problem is relative, as evidenced by the feeling in many cities that five accidents per year at any location is cause for official alarm. There is no acceptable number of accidents, unless it be zero. It is a matter of local decision as to what constitutes a tolerable number of traffic accidents per year and what measures must be employed to reduce the total rate.

Collision Diagrams

Preparation of these diagrams requires complete accident reports for the locations under study, covering all accidents occurring within the last one to five years. The collision diagram shows graphically the nature of all accidents occurring at the location by means of symbols (see fig. 2–6).

The collision diagram need not be drawn to scale but should indicate by arrows the direction of movement of each vehicle or pedestrian involved. The exact point of impact need not be to scale either, although a key number may be added to each collision arrow to provide a cross-reference to the original accident report.

The date and hour of the accident should be written alongside each arrow. Unusual weather or pavement conditions or driver action should also be indicated if it has an important bearing on the accident. If visibility is an important factor in the accident study, night accidents may be circled to provide a comparison between night and day collisions. This same idea may be applied to any single causative factor deserving special recognition.

The collision diagram is particularly useful in revealing accidents falling into one or more of the following classifications, which are clues to needed remedies:

1. Right-angle collisions between vehicles entering on intersecting streets.
2. Left-turn collisions involving vehicles approaching one another.
3. Rear-end collisions.
4. Pedestrian versus vehicle collisions.
5. Collisions between vehicles traveling in the same direction involving turning, lane-changing, or side-swiping.
6. Head-on collisions.
7. Vehicles running off the roadway at curves or constrictions in the roadway.
8. Collisions with fixed objects near the roadside or within the intersection.
9. Collisions with parked vehicles.

The collision diagram may also reveal certain months, days, or hours when most accidents tend to occur, as well as the effect of weather conditions.

Condition Diagrams

Often it is desirable to supplement the collision diagram with a drawing which portrays the existing physical features at, and adjacent to, an intersection or high accident frequency location (see fig. 2–7). This will not always be necessary, however, since physical features of the roadside or intersection may have little bearing on the accident experience. In cases where view obstructions, location, or other factors are suspected, a sketch of physical fea-

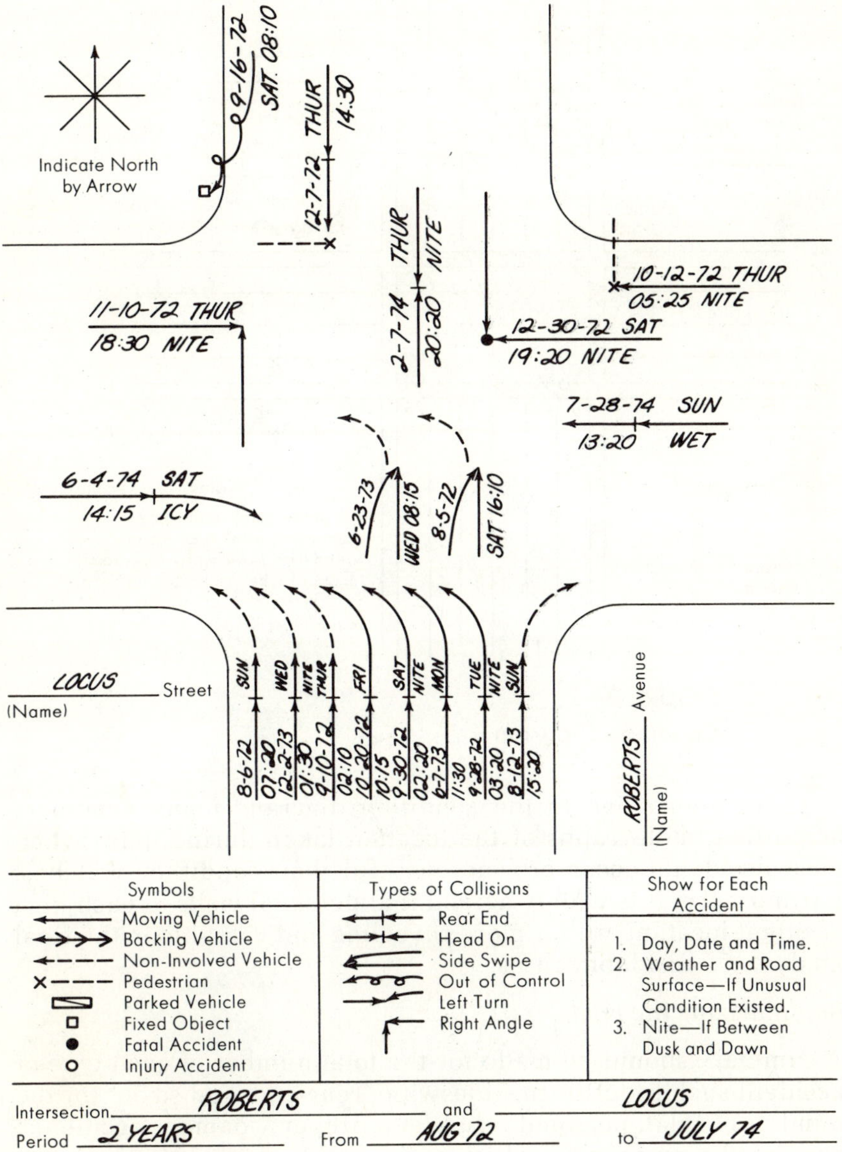

FIGURE 2–6
Collision Diagram

tures should be prepared. Also, it may be possible to add such items to the collision diagram, thus making a combination sketch rather than two separate drawings.

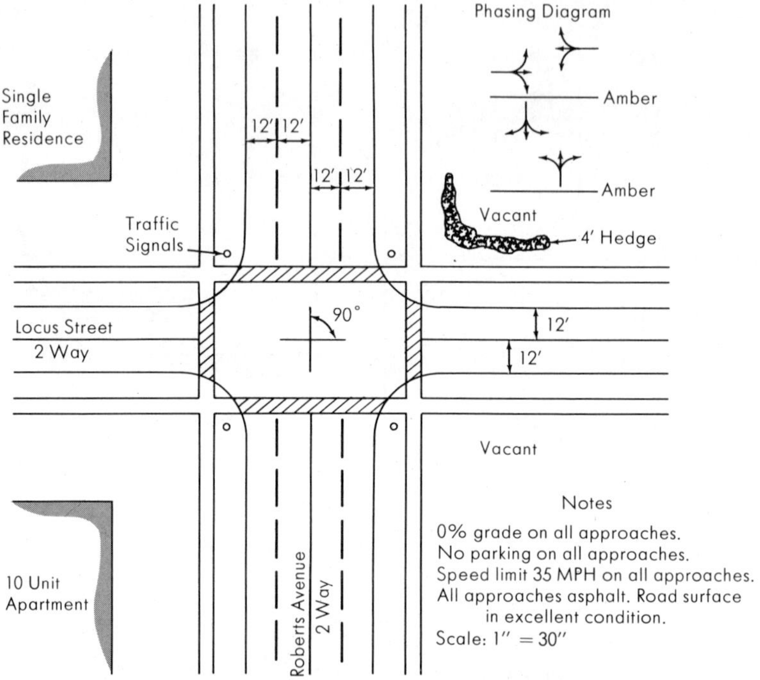

FIGURE 2–7
Condition Diagram

As an alternative to the condition diagram, many engineers agree that photographs of the location taken during times when most accidents occur are more useful than condition sketches. Furthermore, it is widely agreed that personal visits to each high accident location are far more revealing and valuable in the final analysis of accident causes.

Summarizing the Facts

A summary should be made for the total number of each type of accident such as left-turn, sideswipe, rear-end, and so on; for the number of fatal, personal injury and property damage accidents; and for the number of collisions on each approach to the intersection or other location. The numbers of accidents occurring at various seasons of the year and during various periods of the day are also summarized, together with the number of collisions occurring under different weather conditions, such as wet, snow, fog, and dry. Once the summary has been prepared, all outstanding facts should be identified to aid in future references.

Traffic Officer—Traffic Engineer Interrelationships

Both the traffic engineer and the police officer charged with responsibility for traffic control are necessary for the safe movement of traffic in any city, town, or village. The engineer may be a full-time traffic engineer or an engineer of the Public Works or other such department and only serve part-time in the role of traffic engineer. The police officer may be the only member of his or her department specifically responsible for traffic control, or he or she may be a high-ranking officer supervising several thousand officers in coping with the traffic problem of a large city. In any event, neither the police officer nor the traffic engineer alone can provide effective traffic control; one supplements the other. Police seek to prevent accidents under their basic obligation to the community they serve and in order to protect life and property. The traffic engineer studies the characteristic occurrence of accidents because they are a measure of failure in traffic design and operation.

The objective of traffic engineering is to *achieve efficient, free and rapid flow of traffic, yet, at the same time, to prevent traffic accidents and casualties.* The objective of police traffic control is to *secure compliance with the law, expedite the flow of traffic and promote the safety of motorists and pedestrians.*

The basic objective of police traffic control is the movement of traffic with safety. This is also the objective of any good traffic engineer, but engineers are also concerned with the planning and geometric design of streets and highways, and safe, convenient, and economical transportation of persons and goods over these facilities.

The police position is not to usurp the authority of engineers in the broad problems of research, technical design, or construction and the maintenance of streets or highways or traffic control devices. Local conditions relative to the extent of traffic engineering assigned to other city departments will be a controlling factor in the police executive's decision as to what engineering techniques he or she or his or her staff may attempt to perform. However, the scope of police activity in this field should be extended to (1) locating and studying high accident frequency locations, (2) preparing collision and condition diagrams, and (3) developing of accident patterns and simple engineering remedial action.

The one restriction on the role of the police in traffic control is that it stop short of major physical changes in the existing highway and street system. The responsible police official can set in motion through the appropriate authorities such remedial action as signs, signals, or pavement markings where accident experience indicates an intersection or traffic route is congested or accident prone. The role of the police in this field should include channelization and reversible lane procedures as well as establishing alternate routes for use when heavily traveled roads exceed their capacity.

If the police recommendation is not approved by the authority concerned, then it is the duty of the police executive to determine the reasons for such lack of action and to take corrective action under his or her basic obligation to the community to protect life and property therein.

Traffic engineering and control techniques begin with the street and highway system of a community and the characteristics of traffic utilizing such roadways and encompass all the means used to accommodate and control such traffic.

Summary

The purpose of this chapter has been to acquaint the student under his or her basic obligation to the community to protect life engineering so that he or she might understand the importance of law enforcement agencies and the traffic engineer working together toward common objectives to make streets and highways safer and more efficient. The student was presented some of the information needed by a traffic officer to carry out his or her function and to meet the objectives of his or her department. We also attempted to acquaint the student with some of the basic facts about the traffic engineer. The more we learn about traffic engineers and why they do certain things the way they do, the more we are likely to understand and appreciate their problems. On the other hand, an enlightened traffic engineer will become better acquainted with the police officer and his other problems.

If both agencies are to work together, it is essential that mutual understanding and appreciation of each other's problems exist. It is to the advantage of both the traffic engineer and the officer, and above all, the motoring public, if these organizations coordinate their activities so that common objectives are achieved. Traffic engineering is a vast subject and the scope of this chapter has been limited. Later we hope the reader will pursue the subject in depth and become better acquainted with local traffic engineering activities.

DISCUSSION QUESTIONS

1. Discuss why better roads and better automobiles did not supply the ultimate answer to the traffic problem.

2. Define and discuss the role of the highway traffic engineer.

3. Discuss the importance of highway demand and capacity measurements.

4. Discuss the interrelationship between highway capacity and level of service.

5. Discuss the importance of speed zoning in terms of improved highway service.

6. Discuss the primary role of traffic control devices.

7. Discuss the differences, if any, that exist between the role of a traffic officer and a traffic engineer.

3: The Traffic Supervisor

In any organization, the supervisor occupies a position of critical importance. The way an employee values his or her job and even the organization as a whole is colored by his or her relationship with an immediate supervisor.[1] N.F. Iannone describes the role of the supervisor in police work:

> In the law enforcement agency, he [she] is of special importance because of the great need for teamwork. Upon him [her] rests most of the responsibility for providing the cohesive force which welds the working force into a well-functioning, smoothly operating unit.[2]

A traffic supervisor is called upon to perform varied duties, to possess different skills, and to be many things to many people. In this chapter, three areas of basic importance to the traffic supervisor will be discussed in detail: the traffic supervisor as a person; the traffic supervisor as a communicator; and the traffic supervisor as a motivator.

The Traffic Supervisor as a Person

The personality of the supervisor plays an important role in the development of the ability to supervise.[3] While research in this

---

1. Howard M. Carlisle, "The Bond Between You and Your Subordinates," *Supervisory Management* 16, no. 4 (April 1971): 12.
2. N.F. Iannone, *Supervision of Police Personnel*, 2d ed. (Englewood Cliffs, N.J.: Prentice-Hall, 1975), p. 1.
3. Ibid., p. 35.

area makes it clear that there is no such entity as an "ideal supervisory personality,"[4] there are some traits that both research and common sense indicate are of critical importance for a supervisor to possess.[5]

The purpose of this section is (1) to increase the effectiveness of those already in supervisory positions, or aspiring to become supervisors, by helping them concentrate on some specific areas of personality development; and (2) to provide a frame of reference for those administrators and supervisors who are evaluating the suitability of candidates for supervisory positions.

Five personality traits will be discussed which are necessary in an effective traffic supervisor: honesty, trustworthiness, empathy, mistakeability, and assertiveness. Each trait is important and necessary in its own right. Even if a supervisor possesses four out of five qualities, he or she could still experience significant problems. But it also should be kept in mind that personality traits are not "all or nothing" entities. A supervisor could grade himself or herself on the five traits, giving an "A" in trustworthiness, a "C" in assertiveness, and a "D" in mistakeability. So the question is not: "Do I have this trait or not?" but "To what degree do I possess this trait and how can I consciously work to develop it more?"

Honesty

There are two ways in which a person can be honest. One is *morally*, and this deals with having a sound moral character that precludes lying, cheating, stealing, and other forms of misconduct. Obviously, moral honesty is a necessary trait in any police officer, and even more so in a supervisor. But an equally important type of honesty, and one that gets tested much more frequently, is *psychological* honesty.

Psychological honesty means sharing openly, clearly, and consistently what one truly thinks and feels. It means expressing *both* complimentary and critical feelings when they are felt. The supervisor should be an accurate mirror to those he or she supervises so that they can clearly see their own strengths and weak-

4. See Lawrence I. Sank, "Effective and Ineffective Managerial Traits Obtained as Naturalistic Descriptions from Executive Members of a Super-Corporation," *Personnel Psychology* 27, no. 3 (Autumn 1974): 423, and Peter F. Drucker, *The Effective Executive* (New York: Harper & Row, 1967), pp. 21–22.

5. See W.H. Newman and C.E. Summer, Jr., *The Process of Management* (Englewood Cliffs, N.J.: Prentice-Hall, 1961), pp. 220–21, and Bernard M. Bass, *Leadership Psychology and Organizational Behavior* (New York: Harper & Row, 1960), p. 451.

nesses. The supervisor must accept responsibility for his or her own ideas and feelings, e.g., if the sergeant is unhappy with an officer's work, he or she admits to this opinion and does not tell the officer that "The lieutenant is unhappy with your work."

An individual can fail to be honest in a number of ways. One is "selective honesty." The traffic supervisor may share only his or her complimentary thoughts, depriving officers of opportunities to become aware of their weaknesses and the chance to strengthen them. More commonly, the supervisor may express only critical feelings, leaving the officer to wonder if he or she can do anything correctly, and if so, who really cares. The supervisor also may communicate honestly with those who work for him or her, but less so with his or her peers, and almost never to those above him or her.

One basic obstacle to psychological honesty is fear. A person will often be afraid that if he or she is honest, he or she will get hurt or may hurt someone else. The supervisor may fear that if he or she says what he or she truly thinks or feels, some people may not like or respect him or her.

The fear of hurting another person is readily seen in excuses used by a supervisor to avoid being honest with his or her criticisms: "He's got enough problems." "It wouldn't do any good anyway." "It's not the right time." "What if I'm mistaken?" "It will only make matters worse." "It's not that important anyway." "She'll learn sooner or later." None of these rationalizations are valid. The more an officer is allowed to think that things are "fine" when they are not, the more ongoing damage is being perpetrated on him or her. Being "nice" by ignoring a situation that should not be ignored merely postpones facing problems that are incubating each day. A problem that might have been solvable six months ago may now have become unsolvable.

Sometimes the supervisor chooses not to express complimentary feelings because he or she is afraid it may give the officer a "big head," or cause the officer to think the supervisor is getting "soft," or trying to "butter him up." All human beings need positive support as well as criticism. Commending an officer, formally or informally, is likely to have five times the educational effect as a reprimand, even when the reprimand is a valid one.

Despite the obstacles involved, the supervisor should strive to be psychologically honest for three reasons: (1) If a person is emotionally stable and likes and respects himself or herself, he or she should be able to present his or her ideas in such a way that others will like and respect him or her also. This does not mean that everyone will always agree with the supervisor, but disagreement does not preclude admiration or respect.

(2) When the supervisor is dishonest, he or she divides himself or herself into two parts: a real self, which is hidden, and an image which is revealed. It is then the *image* that is liked and respected and not the real person. This should be a hollow satisfaction to any self-respecting individual.

(3) Sooner or later, the supervisor will suffer the consequences of his or her dishonesty. People will come to realize that he or she is a "phony" who is primarily interested in protecting his or her own interests at the expense of fellow workers and the department. The supervisor may also find himself or herself placed in uncomfortable and unhappy situations. For example, a supervisor may allow others to think he or she is more knowledgeable about a particular subject than he or she actually is. On the basis of this image, he or she may be appointed as a teacher at the police academy, a position for which he or she is woefully ill-prepared.

So, while moral honesty is important, it only encompasses all the things that should not be done. Psychological honesty deals with all the things a supervisor can and should do to keep fellow officers abreast of reality.

Trustworthiness

In addition to being competent in his or her work, both academically and experientially, two factors comprise trustworthiness in a supervisor.

(1) He or she keeps *in confidence* all communications that were intended, explicitly or implicitly, to be so. Unfortunately, it takes only one slip-up in this area to undo a supervisor's credibility for a long time.

Why do some supervisors break confidences? Sometimes it is done inadvertently. But this almost never happens with a supervisor who has a reverential attitude toward the privacy of those he or she supervises. He or she protects the privacy of others with the same resoluteness with which he or she protects his or her own.

Sometimes a confidence is broken out of good intentions. The supervisor feels that an officer will be understood better if his or her marital problems are known by a fellow officer. The minor and tentative good that may stem from sharing such information is undone a hundredfold by the damage inflicted on a once trusting and respectful relationship between an officer and his or her supervisor.

As other times, confidential information may be divulged to enliven a gossip session or to make the supervisor look better in one way or another. It should be remembered that when a supervisor violates someone's confidence, he or she is severely damaging two people: the other officer and himself or herself as an effective supervisor.

(2) The trustworthy supervisor is *not hypocritical*. By this is meant that he or she is never "two-faced," saying one thing to one officer and something quite different to another.

For example, Sergeant Smith tells Officer Jones he is doing an outstanding job. But when Jones' partner consults Sergeant Smith about a small problem he is having with Jones, Sergeant Smith shares with the partner his "real concern about Jones not making it through his probationary period." Eventually, the discrepant message filters back to Jones who is both confused and appropriately angry.

People are two-faced for several reasons. If you tell each person what he or she wants to hear, everyone is happy (until they sit down and compare notes). Second, it is a "cheap" way to ventilate one's feelings. The supervisor can tell Smith what upsets him with Jones and vice versa, without ever having to experience the discomfort of facing them directly. And, sometimes, such deception occurs out of malevolence. It is a way of "getting back" at an officer who has been irking the supervisor.

As with broken confidences, deception becomes apparent to those involved sooner or later, and the effectiveness of the supervisor can be severely damaged, if not eliminated.

Empathy

Empathy means a "feeling with" another. It connotes trying to "get inside" another person to see things as he or she does and to feel as he or she is feeling. Empathy is not sympathy, which is feeling pity for another and seldom a healthy sentiment. Empathy does not mean we have to *agree* with how a person perceives or feels about a situation. It simply means we understand the other person's reactions or feelings.

Empathy is important because without this in-depth understanding of another, the supervisor is incapable of communicating with the officer in a meaningful way. In the absence of empathy, the officer becomes a foreign object and in times of conflict is seen only as the source of confusion, anger, and frustration. As long as these feelings persist in the supervisor, there is less chance for a successful resolution to a problem or to a misunderstanding.

Empathy can be a very difficult trait to acquire. Often, if the supervisor is upset with someone, he or she *does not want* to understand him or her; he or she just wants to be upset with that person. Or, the supervisor may be fearful that if he or she really understands the officer, he or she may have to change some of his or her own behavior, attitudes, and beliefs, and that may generate anxiety within the supervisor.

Empathy entails hard work. It takes a lot of listening and gentle questioning. It forgoes easy solutions to problems such as "Don't worry about it"; "I don't care if you understand *why* you are to do it, just *do* it"; "Don't let your partner get to you"; "I'm sure things will turn out all right"; or "Yeah, sure I understand what you mean, but. . . ."

The supervisor will know he or she has reached a point of empathy when, after he or she has paraphrased what the officer has been saying the officer's face lights up and he or she says: "Yeah, now you've got it." It is only then that real understanding and communication can begin.

Mistakeability

Mistakeability is one of the true signs of a mature person. It is the virtue of allowing oneself to make a mistake, accepting the responsibility for that mistake, and admitting it to others, especially to those who were affected by the error.

People who lack mistakeability either deny that they could have made an error; or they admit the error but present "good reasons" that "explain" how it came about; or they take great pains to hide the mistake from the very people who have suffered because of it.

Mistakeability is a rare trait because of two myths. Myth I: If I make a mistake, I am a lesser person. In fact, it is not mistakes that detract from a person but his or her inability to admit to them and rectify them. A supervisor may well make more mistakes than any individual who works for him or her, because the supervisor's role is usually much more complex, stressful, and responsible. Myth II: If I accept and admit my mistakes, I will have to accept those of others, and then everyone will be making mistakes all of the time. In reality, "accepting" mistakes does not connote approval of them. The supervisor does not proclaim: "I made a mistake so you all get to make one now." His or her attitude *should* be: "I made a mistake. Here is how I did it and what I am going to do to insure that I do not make the same error again. I hope we all can learn from this."

This attitude accomplishes two things: (1) it encourages people to admit to their mistakes instead of covering them up (which causes even greater problems at a later date); and (2) sharing the causes of errors can diminish the number of future mistakes because the errors are used as learning experiences for all members of the department.

People who work closely with each other are going to see the others' weaknesses. There are two options. The supervisor and officers can deny their weaknesses, which only adds another great weakness to the list; or they can admit to the flaws, which adds a strength that far overshadows most weaknesses.

Assertiveness

Perhaps the best way to define assertive behavior is first to describe its extremes; nonassertive behavior and aggressive behavior. The nonassertive individual is often shy, timid, and reserved. He or she finds it very difficult to stand up for his or her rights and to act on his or her feelings. He or she easily gives in to any request made of him or her or feels guilty for turning someone down. When things go wrong he or she blames himself or herself first.

As a supervisor, this individual experiences many problems. He or she is taken advantage of by both those above him or her and those below him or her. He or she lacks the strength to "back" his or her officers in conflicts with citizens and superior officers. He or she encourages them to "forget" injustices, which creates resentment among those he or she supervises. He or she and the people he or she supervises are passed over for benefits time and again because he or she fails to let the needs and hopes of his or her officers be known to those who could help them.

On the other end of the scale is the aggressive person. He or she "puts himself or herself up" by "putting others down." The chief aim of the aggressive person is to get what he or she wants, no matter what others have to pay in terms of time, energy, money, pressure, or personal hurt.

There are two ways to be aggressive. One is on a physical level. This type of individual dominates his or her environment and demonstrates his or her toughness with bravado. But a more intellectually oriented person can be equally aggressive. He or she may dominate conversation—both by verbosity and loudness—belittle the opinions of others, and leave no doubt that he or she considers himself or herself the final word on nearly every topic. He or she often confuses managing people's *work* with managing their *humanity*. Often this individual does not wish to be aggressive, but he or she has not learned responses that are appropriately assertive.

As a supervisor, the aggressive person is a threat to the people around him or her. People feel insecure in his or her presence because they realize that their best interests are not his or her concern. They feel, rightly, that they are expendable and dare not express to the supervisor's face what they say and feel behind his or her back. As a result, fear reigns and morale is low.

At a midpoint between the nonassertive and the aggressive supervisor is the assertive supervisor. This person is able to let others know who he or she is and what he or she wants without damanging people in the process. He or she respects his or her subordinates' human rights just as conscientiously as he or she expects others to respect his or hers. He or she is fully in charge of himself or herself in interpersonal relationships, displaying confidence and capability without arrogance or hostility. He or she realizes that as a supervisor he or she has no right to take advantage of his or her subordinates' natural rights to courtesy and respect as human beings.[6]

As a supervisor, this individual is respected because his or her respect for others is so continual and obvious. His or her ability to approach people openly and honestly invites others to approach him or her on the same basis. People feel safe around him or her because they know he or she operates in ways that will be in their best interest as well as his or her own and those of the department.

The Supervisor as Communicator

"For supervisors—the strength of an organization and its real men and women in the middle—personalized communication should be a primary professional goal" states John Brennan.[7] A large percentage of the problems in any organization can be traced

6. Robert E. Alberti and Michael L. Emmons, *Your Perfect Right* (San Luis Obispo, Calif.: Impact, 1974), pp. 24–25.

7. John Brennan, *The Conscious Communicator* (Menlo Park, Calif.: Addison-Wesley, 1974), p. 190.

eventually to faulty communication.[8] The price an organization pays for ineffective communication can be measured objectively in number of dollars lost and employee hours wasted.

A person is not, by nature, a good communicator or a poor communicator. Good communicators are people who have taken the time to learn communication skills and to practice them. Effective communication depends on skills that can be learned. This fact was brought out sharply by a study in which police officers trained in reporting their observations did better than Phi Beta Kappa college seniors who had no such training.[9]

Because communication is the supervisor's most important, yet most difficult responsiblity, a number of factors in the communication process will be discussed. The principles noted in this section are, for the most part, valid for all types of communication. However, the emphasis will be on verbal communication. This is because face-to-face communication is superior, under most circumstances, to written orders, memos, or business letters.[10]

Effective Communication Means Listening

Listening is perhaps the most overlooked aspect of the communication process. The word "communicate" almost always connotes talking rather than listening. Yet, no matter how eloquently a person speaks, if no one is listening to him or her, no communication is taking place. As Roy G. Foltz states: "Everyone who communicates should spend at least half of his time listening, and professional communicators should spend even a higher percentage of time listening, particularly to those things that are emotionally upsetting."[11]

To become an effective listener, the traffic supervisor should be aware of some of the barriers to effective listening and some of the aids that can help him or her become a more active listener. Some barriers to effective listening are:

1. *Internal noise.* This term indicates that there is so much inner "static" going on within the individual that it significantly drowns out the bulk of what is being said to him or her. This inner noise is typically caused either by anxiety or anger. The listener is

8. Roy G. Foltz, "Communication: Not an Art, a Necessity," *Personnel* 49, no. 3 (May-June 1972): 63.
9. Frank W. Draden and John T. Trutter, "Why Communication Goes Haywire," *Supervisory Management* 12, no. 1 (January 1967): 10.
10. Leonard R. Sayles and George Strauss, *Human Behavior in Organizations* (Englewood Cliffs, N.J.: Prentice-Hall, 1966), p. 249.
11. Foltz, "Communication," p. 62.

so threatened or angered by the situation (either the content of the message, or the sender of the message, or both) that he or she "cannot hear."

The astute supervisor perceives this officer's "tuned-out" look, the fact that he or she asks questions that have been answered, or makes irrelevant statements. The supervisor then chooses to discontinue his or her primary message in order to deal with the "hidden agenda," that is, the smoldering feelings of the officer. Once the feelings are ventilated, there will be a more open receptivity to the original message. If the supervisor is experiencing internal noise, he or she should deal with it before attempting to continue the conversation.

2. *Judgmental listening.* Passing premature judgments, whether they are favorable or not, tends to restrict an employee's freedom of expression. Even a poorly timed positive evaluation can be inhibiting because it could make it difficult for the person supervised to talk about mistakes he or she has made. Premature criticism may ultimately turn out to be unjust and, in any case, can discourage the officer from continuing with additional parts of his or her story. A good listener sends the message: "Tell me *everything,* and I will react after you let me know you are completely finished."

3. *An authoritarian attitude.* If the supervisor feels that "What I have to say is more important than what you have to say," he or she will be formulating his or her next words instead of listening. Or, he or she may convey the message: "Tell me only what I want to hear." This results in a filtering or cover-up of real problems. Often, the information that the supervisor is protected from will eventually return to haunt him or her.

4. *Misunderstanding silences.* A supervisor who is a good listener realizes that when the other person stops speaking, it does not necessarily mean that he or she has stopped communicating. Often, a silence is a part of an ongoing communication during which the speaker is reevaluating to himself or herself what he or she has just said, or formulating in his or her mind how he or she will present the remainder of his or her message. To interrupt at that time may demolish the speaker's train of thought and emotional state. It is not when the speaker's mouth stops moving that conveys when he or she is finished speaking; it is when his or her eyes "rest" and say: "OK, now it is your turn."

What are some aids to effective listening?

1. *Listen actively.* The supervisor's whole demeanor should re-

flect that he or she is actively involved in the listening process. His or her attitude, posture, eyes, and facial expression should make it clear to the officer that he or she is "with" him or her all the way. The actively listening supervisor asks questions that draw out the officer's ideas and feelings. The supervisor also checks up on what he or she has heard by restating it the way he or she understands it to validate his or her perceptions.

If the supervisor's mind wanders, he or she feels free to say: "I'm sorry, I was distracted for a second, could you repeat that?" He or she encourages the officer by nodding his or her head and periodically reassuring him or her by saying, "I understand"; or "I could see how you would feel that way." If the supervisor senses some confusion or contradiction in the officer's story, he or she patiently asks for clarification, e.g., "I think I missed a step back there," versus "Hey, wait a minute; you just contradicted yourself!"

2. *Listen with the "third ear."* A good listener has the ability to go beyond the spoken words to their *meaning* and *feeling* levels. For example, an officer may say to his supervisor: "I really need to get off 'swing shift'; it's creating havoc with my life."

The poor listener who is merely hearing words and not messages replies: "I'm sorry, Bill, but you have to take your turn with 'swing' along with everyone else—it's part of the game."

The supervisor who is an astute listener is aware of the possibility of other messages:

—Is Bill trying to tell me that he and his wife are having problems?

—Is he telling me he is under a great deal of psychological stress and that he is becoming worried about it?

—Are his kids developing behavioral problems because he does not see them much anymore?

—Is there something going on during the swing shift that is upsetting him?

These are all areas that a good supervisor will want to gently explore before making a final decision.

3. *Learn to read nonverbal communication.* Someone once said that the "loudest" kind of communication is the kind we cannot hear. What this means is that an important part of "listening" is done with one's eyes. Facial expressions, gestures, posture, and mannerisms may transmit a message that is not being sent verbally, or may contradict a message that is being sent verbally. A smirk on someone's face, clenched fists, finger tapping, wander-

ing eyes, looks of boredom or upset, slouched, or rigid posture, a smile, all may express messages that a good listener will perceive and pursue in a friendly way. A poor listener will either miss these nonverbal cues or purposely ignore them because he or she does not wish to "hear" what the messages are saying. As Bernard B. Beegle states:

> It is the supervisor's own interest to *observe* how his [her] subordinates communicate through gestures and mannerisms and to acknowledge these signs as legitimate messages that demand attention. When employees communicate non-verbally, they are signalling you to get involved. They are asking you to take the initiative to talk to them, to find out if problems exist, and to take action to solve them quickly.[12]

4. *Encourage honest feedback.* All three of these words are important. "Encourage" means that the supervisor, in a clear and nonthreatening manner, invites the officer to share his or her ideas and feelings. The supervisor's message is: "I'm really interested in what your feelings are on this—what you think about it is important to me." This is contrary to the supervisor whose message is: "Well, if there is nothing else you may return to work."

"Honest" means that the supervisor conveys a sense of security that assures the officer that he or she can "take in stride" any reaction that the officer may offer. The message is: "There may be some things you agree with and don't agree with. I need to hear both because I want to understand what this means to you." This is in contrast to the supervisor who conveys: "Well, what I said makes pretty good sense, don't you think?"

Feedback, in itself, is important. Without it, the supervisor is deprived of information he needs to make good decisions. It also deprives the officer of feeling that his or her ideas are important enough to be considered in the decision-making process.

Listening pays off in three ways. The supervisor gets all the facts and feelings from his or her officers so that he or she feels confident of the situation. Second, the officer feels that the supervisor really cares about him or her. Finally, the officer is now willing to listen with openness to what the supervisor would like to say.

---

12. Bernard B. Beegle, "The Message That Is Sent Without Words," *Supervisory Management* 16, no. 2 (February 1971): 14.

Effective Communication Is Direct

When a supervisor wishes to share some information with an officer, be it a commendation, reprimand, or just general information, he or she should communicate it directly, preferably in a verbal message, but at least with a written communication. This seems simple enough, yet it is one of the biggest causes of failure in communication.

Being *indirect* is often easier. One can be indirect by communicating "by inference." A supervisor protests: "Why should I have to *tell* Officer Smith she is doing a good job. If she hasn't been 'called on the carpet' recently, she should *know* she is doing well!" Or, "If Officer Smith can't tell she's got problems around here by the way I treat her, she's beyond hope anyway." In both instances, if the supervisor spent ten minutes with the officer, he or she could undoubtedly save hours of time at some later date.

The supervisor can be indirect by being subtle. Subtlety is employed mostly with unpleasant communications. The supervisor "drops hints," veiled suggestions, and innuendos that the officer's performance or behavior is unsatisfactory. The supervisor then says: "I hope he got the point." But there is no place for "decoding" messages in healthy communication. If the officer failed to "catch" the point, it is because the supervisor failed to *make* the point.

A third way of being indirect is for the supervisor to express his or her thoughts about an officer to a third party, hoping that the "word" will filter down to the officer. This is almost always disasterous because the message is often contaminated by the messengers, depending upon their involvement in the situation. The result is that, by the time the communication finally gets to the intended officer, it is often distorted in some way. This, in turn, creates a new set of problems.

Effective Communication Is Immediate

As little time as possible should elapse between feeling the need to communicate and the actual communication. Immediacy is important for two reasons. If the situation calls for the expression of unpleasant feelings (anger, disappointment, criticism), the longer the supervisor holds in these feelings, the more likely they will be overexpressed (because they incubate), underexpressed (because they wear off), or never expressed. In none of these cases will the officer benefit from knowing how his or her supervisor truly feels; hence, he or she may overadjust his or her behavior, underadjust it, or fail to adjust it.

Second, immediacy is important because the closer in time a person is rewarded or punished for his or her behavior, the stronger the linkage between the behavior and the consequences. This is essential for true learning to occur.

The two main barriers to immediacy in communication are procrastination and disturbed priorities. People often put off what will create anxiety within them; they are chronically awaiting the "right time" to convey the message. But in effective communication, there is only *one* right time, and that is the present. Disturbed priorities are evident when a supervisor is "too busy" to communicate issues while they are still "hot." He or she has letters to write and meetings to attend, which he or she thinks are more important than communicating with subordinates. The more a supervisor interposes other duties between himself or herself and those he or she supervises, the more distance there will be between them.

Effective Communication Is Clear

Communication should be concise and lucid. Most people talk too much. This is done to camouflage their insecurity stemming from not knowing what they really want to say, or knowing but being reluctant to "get to the point." It is often helpful for a supervisor to phrase mentally the one or two thoughts that he or she wishes to convey. Then, when the time comes, the supervisor is better prepared to share those ideas in a simple and concise manner. The more words that are used, and the more decorative they are, the more cause for distraction, inattention, and confusion.

There are three common types of unclear messages. One is the ambiguous message. The supervisor tells an officer: "I want you to concentrate on the speeding drivers on Main Street, but I also want you to use your head." What does this mean?

An unclear message is an inconsistent one. A supervisor may tell his or her officers: "I don't want any more complaints from citizens about your hiding to catch drivers 'blowing' stop signs." Two weeks later the same supervisor tells the same officers: "You people are going to have to start enforcing stop signs. We've had three injury accidents this month at those intersections."

Third, an unclear message is a "double message." A supervisor says: "I want you to enforce the double-parking laws downtown, but I don't want the merchants up here complaining." These are "damned if you do" and "damned if you don't" messages.

A good communicator is secure enough in what he or she thinks and feels that he or she can afford to be to the point and lucid without being blunt.

Effective Communication Is "Pure"

A "pure" communication is one in which the *stated* purpose for the communication and the *real* purpose are identical. A "contaminated" communication occurs when there is a discrepancy between the real and stated purposes.

For example, a supervisor may advise an officer: "I'd like to talk to you about a few problems with your accident reports." If the communication is "pure," this is exactly what will happen. But perhaps the supervisor is using the officer's traffic reports as a pseudo-issue. The supervisor may feel that the officer has been "goldbricking" lately and is angry with him. So, while the supervisor's purported motive is to help the officer with his reports, the real purpose is to "stick it" to the officer.

The results of this contaminated communication are:

(1) The officer is confused because he *hears* what the supervisor is saying about the reports but he *feels* waves of hostility and punitiveness which seem incongruous and inappropriate.

(2) The officer may feel the supervisor's wrath but leave the discussion not knowing the cause of his anger. Consequently, he will be unable to make appropriate changes in his behavior.

Other "hidden motives" can be seen in the following "double-plays."

—I want to talk with you about your reports (but I really want to impress you that I know more about your work than you do).

—I want to talk with you about your reports (but I really want you to know that I am your boss which you do not seem to realize).

—I want to talk with you about your reports (but I really want to show you what a nice person I am so you will like me).

Most hidden motives could be legitimate issues for communication, but they should be dealt with as separate communications and not camouflaged by other issues.

Effective Communication Is Constructive

"Constructive" communication means that the purpose of the communication is to better the other person, not to tear him or her down or to destroy him or her. The message in a communication can be viewed, analogously, as a brick. A brick can be *thrown* at another person, but this will cause a "fight or flight" reaction in him or her. Either he or she will fight, i.e., throw another brick; or he or she will flee—neither of which creates an atmosphere for effective communication.

But a brick can also be used to build a bridge between the supervisor and the officer. A supervisor can express displeasure, disappointment, or frustration in a way that reflects on the individual's particular behavior but not on who he or she is as a person. For example, a supervisor may state: "I noticed that you have been late for our last three meetings; you seem like an irresponsible person."

"Throwing bricks" can be done in several ways. One is by the vocabulary we use. Referring to people with emotionally loaded labels such as "dumb," "lazy," "irresponsible," or "dishonest" accomplishes nothing except to increase the listener's defenses. The more defenses that are erected, the less real listening and thoughtful consideration will be given to what is being said.

"Brick throwing" can also be done by the use of sarcasm which is a very ineffectual method for communicating. Sarcasm is belittling and hostile, and it precludes mutual openness which is a prime requisite for effective communication.

Bringing up old or unrelated issues is another "brick." This is often prefaced by "Now that we are getting things off our chest . . . ." If the side issues are important, they should have been brought up when they were "hot"; if they are not as important, they should not be added to the main issue.

Embarrassing another person in front of others is also a situation that interferes with communication. The message gets lost in the inner static that is created by the feeling of humiliation the listener has.

Comparisons also are likely to interfere rather than to help the reception of a message. Suggestions such as: "Why don't you watch Officer Smith; *she* knows what she is doing" are seldom helpful. The main effect of such a statement is that the officer builds up a resentment toward *both* the supervisor *and* officer Smith.

Making inferences about behavior rather than focusing on observations can be destructive. For example, it is better to say: "I noticed you had a hard time keeping awake at the meeting, and I was wondering what it meant," than to say: "You were obviously bored stiff at the meeting—what's the problem?"

Attributing motives to an officer's behavior can be problematic as well. It is more constructive to say: "You seem to be having trouble with your accident investigations," than to accuse: "I suppose you think that if you create enough problems in accident investigation, I'll let you go back to routine traffic patrol."

Even the most severe criticism can be imparted without angry and denigrating words. The question the supervisor can ask himself or herself is whether his or her aim is to destroy the individual or open up channels for communication.

In constructive communication, the main attitude of the supervisor should be one of support. One method of accomplishing this is to point out concrete examples of where the officer handled something well, so that the mistake may be put into proper perspective.

Another way to be supportive is for the supervisor to share the fact that he or she may have made the same or comparable mistakes in the past and to elucidate the negative consequences that flowed from it. The officer who leaves a discussion with his or her supervisor feeling "I blew it" while still respecting the supervisor as a person and an officer is infinitely more likely to rectify a mistake than the officer who leaves feeling "He doesn't respect me anymore. Well, I don't care *what* he thinks." When mistakes are pointed out in an atmosphere of realistic support and encouragement, better cooperation and more effective behavior will result in the officer.

The Traffic Supervisor as Motivator

"A supervisor's essential task is to create a climate in which employees satisfy their own needs by working to fulfill department objectives" states Howard P. Smith.[13] In attempting to accomplish this task, the supervisor should realize three points:

(1) In the last quarter century, it has become obvious that money (salary, financial rewards and benefits) is no longer a significant motivation for most people.[14] Since individuals have become more educated and sophisticated, it is increasingly important for management to fulfill their *emotional* needs.[15]

(2) Motivation, then, is not something that a supervisor *does* to his or her workers.[16] A supervisor cannot give motivation to his or her people. He or she can only create a work situation in

---

13. Howard P. Smith, "Keys to Employee Motivation," *Supervisory Management* 16, no. 2 (February 1971): 41.
14. Clois E. Bristo, "Do You Motivate Your Subordinates?" *Supervisory Management* 16, no. 9 (September 1971): 12.
15. Vincent W. Kafka, "A Motivation System That Works Both Ways," *Personnel* 49, no. 4 (July-August 1972): 62.
16. John R. Hinrichs, *The Motivation Crisis: Winding Down and Turning Off* (New York: Amacom, 1974), p. 44.

which the workers can get their needs met both through the work they do and through their relationships with the supervisor.

(3) There are two main reasons for a supervisor being committed to the fulfillment of his or her subordinates' emotional needs: (a) On a humanitarian level, it is intrinsically good to offer other human beings the opportunity for growth and happiness. (b) On a pragmatic level, a person who is getting good "emotional groceries" at work will concomittantly develop good "psychological muscles." This enables him or her to rise above many of the conflicts that can plague an organization: petty jealousies and rivalries; suspiciousness that others are against him or her or getting more than he or she; gossip and backbiting; and a lethargic, uncommitted approach to his or her work. This officer is more resilient to stress, more adaptable, and more interested in the goals of the department.[17]

The Nature of Emotional Needs

The chief goal of the supervisor is to provide opportunities for the employee to *move* from being a less satisfied to a more satisfied human being, and, consequently, to being a more effective worker. We will discuss some of the theoretical and practical principles involved in emotional growth. We do not intend to provide a "cookbook" to be followed literally, but rather to present a frame of reference for understanding and relating with people in a work situation.

There are three categories of needs: maintenance needs, growth needs, and self-actualization needs.[18] These three categories represent a hierarchy, with maintenance needs at the base (see fig. 3-1). Once maintenance needs are met adequately, the person progresses to growth needs. When these are satisfactorily fulfilled, he or she then "graduates" to self-actualization needs. Because this hierarchy represents a growth process, there may be some overlapping between one level of need and another. When an individual progresses from a lower level need to a higher one, the lower need does not disappear. It will continue to exist but its strength and importance will be replaced by the higher level need. For example, a self-actualized individual still needs to feel safe, accepted, and esteemed. But his or her main source of motivation stems from higher needs.

---

17. Saul W. Gellerman, *Management by Motivation* (New York: American Management Association, 1968), p. 50.

18. See F. Herzberg, B. Mausner, and B.B. Snyderman, *The Motivation to Work* (New York: John Wiley, 1959), and Abraham H. Maslow, *Toward a Psychology of Being* (New York: Van Nostrand, 1962).

Maintenance Needs. This first level is comprised of two sub-levels: the first is physical security (the desire for food, water, warmth, rest, health). When this stratum of needs is adequately met, the focus progresses to emotional security (money, working conditions, freedom from arbitrary decisions, freedom from threatening relationships with fellow workers or supervisors, job security). Research indicates that when maintenance needs are not met, workers are dissatisfied, and this creates poor morale and lowered production. *But, when maintenance needs are met, this does not produce job satisfaction—it only prevents dissatisfaction.*[19]

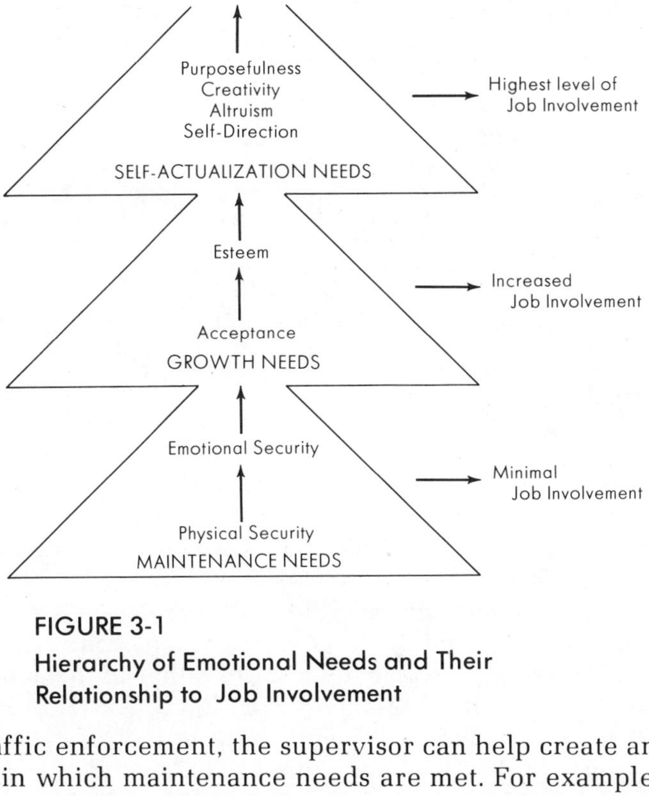

FIGURE 3-1
Hierarchy of Emotional Needs and Their Relationship to Job Involvement

In traffic enforcement, the supervisor can help create an atmosphere in which maintenance needs are met. For example, he or she can see that work schedules are reasonable and create a minimum of hardship on personal health and family life. The supervisor can insure that the officer's vehicle and emergency equipment are properly maintained so as to avoid any unsafe or embarrassing situations for the officer.

19. Hinrichs, *The Motivation Crisis,* pp. 91, 139–40.

The supervisor can help dissatisfied partners work out their differences or separate them. The good supervisor does not threaten his or her officers' emotional security by spying on them, withholding information from them, embarrassing them publicly, or threatening them with anger, unreasonableness, or power. He or she conveys the message: "I will do all I can to free you from 'nuisance worries' so that you can concentrate on the job you were hired to do."

Growth Needs. Once maintenance need fulfillment can be taken for granted, growth needs become salient. Growth needs are also comprised of two sublevels: the need to be accepted and the need for esteem.

The need to be accepted includes the desire to belong and to be liked, affirmed, supported, and appreciated. One important way the supervisor can contribute here is by relating *with* his or her subordinates and not *to* them. As Saul Gellerman states: "The most effective supervisor is a catalyst, not a drill sergeant. . . . The self-important, self-impressed supervisor with a passion for keeping his 'prerogatives' to himself and his subordinates 'in their place' may still be commonplace, but he is also obsolete.[20]

The good supervisor can allow his or her subordinates to feel they are friends without fearing that they will take advantage of his or her friendship. He or she realizes that being a friend also means being honest, strong, guiding, and, at times, corrective.

Supervisory attitudes that short-circuit the fulfillment of acceptance needs include:

—"I was hoping you wouldn't be transferred to my department, but now that you're here, let's make the best of it."

—"You're on your own, and God help you if you mess things up."

—"We can get along fine without you."

These attitudes are quite human and, at times, may even be appropriate. However, they are not helpful either to the supervisor or to the officer. Either the supervisor should try to understand better his or her own feelings and those of the officer to see if some rapport can be reached, or he or she should see to it that the officer is transferred.

Messages that convey acceptance are:

—"I'm glad I have you in my department."

---

20. Gellerman, *Management by Motivation*, p. 39.

—"I like you as a *person*, not just someone who issues a lot of traffic citations and investigates a lot of accidents."

—"When you are experiencing difficulty with anything, feel free to bring it up, and we'll work on it together."

—"I really appreciate the extra work you did for us the other day."

It is necessary to realize that *saying* these things is more helpful than not saying them. But, more importantly, they should not be just statements but sentiments; not manipulative devices, but genuine reflections of the supervisor's feelings and attitudes.

Esteem needs follow the successful fulfillment of acceptance needs. Esteem needs include the desire to achieve and have one's achievements recognized; to feel skilled, important, and rewarded. Officers should be allowed to achieve as far as their talents will take them. To assign a person with a new bachelor's degree in administration of justice to marking tires (putting chalk marks on tires to determine if the vehicle has been moved over a period of time) for overtime parking may be "starting him or her at the bottom, so he or she will know how it feels." But, if that is his or her eight hours a day job for more than a few months, law enforcement will lose a good individual. Interested and talented officers can be encouraged to address service organizations, to give seminars both within the department and outside of it, and to publish and to teach.

A supervisor cannot *make* a person feel important. He or she only can allow the officer to *be* important. One way of doing this is to allow the officer to participate in the decision-making processes that effect him or her. Such individuals perform at a higher level than those who are not given such an opportunity.[21]

The supervisor can provide nonmaterial rewards. This includes allowing the officer to take on more responsibility; genuinely asking him or her for advice in an area with which he or she is particularly familiar; asking him or her to help out the supervisor on a particularly important project; and giving him or her full credit for his or her accomplishments. This attitude is in contrast to one that says: "You're doing a good job; aren't you glad you had me to teach you all these things?"

When the officer's growth needs are being met, the organization as a whole should benefit. As Adams states: "The employee is more likely to produce a better quantity and quality of service if

21. Victor H. Vroom, *Work and Motivation* (New York: John Wiley, 1964), p. 267.

he feels that he is understood, accepted, liked, and that his work is generally approved." [22]

Self-Actualization Needs. When a person's growth needs have been met well for a good length of time, he or she transends to the highest and infinite level, that of self-actualization needs. Among the more important self-actualization needs are self-direction, altruism, creativity, and the need to see a meaningful purpose in one's work.

The need for self-direction means a desire to be more self-determining and autonomous. It is manifested in the wish to make more decisions, to become individually responsible for one's work and more "on one's own." This need does not flow from needing other persons less but from a growth in self-confidence and an increased willingness to trust oneself and to accept the consequences for one's own actions.

Altruism or the need to give of oneself unconditionally indicates that the person wishes to give for the sake of giving, and not for ulterior motives (as might be seen in the need for acceptance). This officer enjoys teaching younger officers and sharing his or her experiences with them. He or she gives to the department and to the community in ways for which he or she could never be adequately rewarded.

The need to be creative is seen in the attempt to solve problems using new and unique approaches. The individual is willing to approach a situation from a new light, sometimes risking ridicule or failure.

Because of the need to find a meaningful purpose in one's work, the old reasons for working (money, fellowship, prestige) are still present but are no longer as compelling. The work itself has to make some sense; it has to have some perceived inherent value.

The officer with self-actualization needs is the greatest challenge to the department. It takes a self-actualized administration and supervisorial staff to provide adequate opportunities for this officer's growth. It means allowing the officer a high degree of self-direction instead of keeping him or her "under thumb." It requires providing opportunities for the officer to give of himself or herself instead of restricting him or her to routine assignments. The officer should be given opportunities to try new approaches and allowed to enjoy his or her successes and learn from his or her failures. The supervisor must allow the officer to pursue

22. Thomas F. Adams, Gerald Buck, and Don Hallstrom, *Criminal Justice Organization and Management* (Pacific Palisades, Calif.: Goodyear, 1974), p. 129.

facets of his or her work that are intrinsically meaningful to him or her in contrast to performing tasks merely to receive a paycheck, or to get a pat on the back or a promotion. The officer has to feel that a little piece of the world is going to be slightly better because of the work he or she performs.

There are a number of ideas that a supervisor should be aware of with regard to allowing his or her workers to meet their emotional needs.

(1) The supervisor is not solely responsible for the emotional growth of his or her subordinates. The officer's family, friends, church, and social organizations also are influences. Moreover, the supervisor cannot assume responsibility for the total work situation. There are some factors over which he or she has little or no control, such as fiscal matters and major policy pronouncements. However, the supervisor should recognize that he or she is one of the most important influences in the officer's life. Hence, the supervisor has ample opportunity to create an atmosphere in which the officer can grow, stagnate, or regress.

(2) The supervisor is in no way asked to sacrifice his or her own need fulfillment for those of his or her subordinates. What was said about the critical importance of need satisfaction applies to supervisors as well. If a supervisor is not seeing to it that his or her emotional needs are being sufficiently met at work, he or she will be unable or reluctant to meet those of others.

(3) The supervisor can realize that once an officer's needs on any particular level are reasonably satisfied, they no longer act as a motivator. For example, giving an officer a chance to make overtime pay (security need) is going to be ineffectual as a motivator if what the officer *emotionally* needs is to feel a sense of achievement in his or her work (growth need).

When this fact is not understood it creates frustration for both the supervisor and the officer. The officer feels misunderstood. The supervisor feels confused because he or she has gone out of his or her way to do a favor for the officer who does not seem to be grateful.

So, a supervisor must know, at least generally, what each individual needs in order to provide appropriate opportunities for growth. He or she must realize that what motivates Smith does not motivate Jones and to confuse the needs might result in negative consequences. He or she must also realize that what motivated Jones two years ago (teaching recruits, an esteem need) has been traded in for a higher need (creating new procedures for accident investigation—a self-actualization need). Self-actualization needs, however, are infinite and are never totally met.

(4) It is helpful for a supervisor to understand this relationship between complaints from his or her subordinates and their level of need fulfillment. The supervisor who dreams of the day when no one complains to or about him or her is dreaming of the day when he or she is supervising angels and not human beings.

Complaints should not be viewed as criticisms to be defended but as clues to be interpreted. Psychologist Abraham Maslow writes:

> . . . we should, according to motivation theory, never expect a cessation of complaints; we should expect only that these complaints will get to be higher and higher complaints, i.e., that they will move from the lower-grumble level to higher-grumble levels and finally to metagrumble levels.[23]

Hence, as a supervisor is more successful in allowing officers to motivate themselves, their complaints will not diminish necessarily, but they will be of a higher quality. An officer may begin employment with his or her complaints centering around topics such as the fact that the highway flares issued to him or her were of cheap quality. Five years later the same officer may be complaining about not being given complete freedom in running his or her accident investigation team. If this is so, the supervisor can feel that both he or she and the officer are growing in their job situation. The nature of the complaint, then, is often a solid clue as to what level of need fulfillment the officer has reached and what he or she needs next if he or she is going to continue to grow.

Summary

In summary, the traffic supervisor is the fulcrum of the traffic department. On one side is the administration, expecting him or her to meet their needs. On the other side are the line officers, expecting him or her to meet *their* needs. And, as a healthy human being, the supervisor is also trying to meet *his* or *her own* needs. He or she will never be able to always keep "all three balls" in the air. At one time or another, he or she is going to let someone down. He or she should expect that periodically this will occur, and he or she should neither berate himself or herself for this nor blame others for expecting too much of him or her. This is merely a fact of life for supervisors.

A supervisor can significantly increase his or her effectiveness and decrease his or her chances for failure by trying to acquire

23. Abraham Maslow, *The Farther Reaches of Human Nature* (New York: Viking, 1971), p. 242.

some of the personality traits of a good supervisor. These are not inherent; they are learned. As in learning anything, it takes awareness of one's behavior and daily practice.

A supervisor can also learn to communicate in an effective way. He or she should realize that, as a supervisor, he or she is *always* communicating. The question is whether what he or she is communicating is a destructive or constructive message.

The supervisor can understand what the true nature of motivation means. He or she must realize that his or her role is not to infuse motivation but to allow people to unleash their own intrinsic motivation.

So, the key to success for the supervisor is understanding human behavior and relating personally with people. As Harold Mayfield says of the supervisor:

> He [she] establishes the climate of the workplace for the people under him [her]. Simply by acknowledging that his [her] workers are human beings and by treating them considerately, a supervisor can do much to erase the notion from their minds that large organizations are impersonal and inhuman. His [her] hands are not often tied by higher authority when it comes to handling his [her] personal relationships with subordinates. And rarely will someone higher up in the organization make up for his [her] deficiencies in this respect. This is his [her] responsibility.[24]

DISCUSSION QUESTIONS

1. "*Who* a supervisor *is* is often more important than what he does." What does this statement mean and what are some of its practical implications?

2. Of the five traits discussed in this chapter (honesty, trustworthiness, empathy, mistakeability, and assertiveness), which is the most difficult for you to "master"? Why is this so? What can you do to develop this trait?

3. In communication, listening is very likely the most difficult part of the process. Why is something that should be so very easy, so terribly difficult?

4. Comment on the statement: "The question is not whether a supervisor is communicating or not, since he is *always* communicating; the question is, what is he communicating?"

---

24. Harold Mayfield, "Who Is 'The Company'?" *Supervisory Management* 12, no. 3 (March 1967): 15.

5. Comment on the statement: "If you tell people they are doing a good job, or tell them that you appreciate them, or tell them how important they are—this should adequately meet their needs."

4: Traffic Records Systems

A traffic records system is the collection of documents, files, procedures, and summaries that provide a history of traffic problems in a jurisdiction in a way that permits assessment of overall cause and effect, that is, causes and effects of the traffic problem as a whole. It includes, but is far broader than, a system that embraces officer procedures for documenting accidents and traffic violations and the subsequent indexing and filing of these documents by the Records Bureau.

While the mechanics of report preparation and filing are vital to a traffic function, these considerations are appropriately handled by the many good texts available today on police administration. This chapter will thus examine traffic records systems from the supervisor's perspective. The emphasis will be on traffic records systems as management tools, and how overall improvement of the traffic problem can be enhanced with a traffic records system.

Purpose

The purpose of a traffic records system is to provide informational feedback of the traffic supervisor's plans and actions to improve traffic control. A traffic records system will prove for the supervisor how well his or her methods are working, and where further effort may be most advantageously applied.

For example, it has been demonstrated that the typical traffic accident has a total cost to society of $2,800.[1] A traffic records system that shows a traffic supervisor how he or she can prevent an accident a week has a potential benefit of over $140,000 in one year! Or, if the supervisor could discover ways to prevent an accident a day, the potential benefit would exceed $1 million annually!

The traffic records system's major benefit is for planning activities. This planning can be directed toward preventing accidents and violations and to optimally allocate patrol resources so as to provide uniform and effective enforcement of traffic laws and response to traffic emergencies. A traffic records system provides the data base for studies of accidents and citations and, when automated, the capability for data handling necessary to reveal underlying trends.

Overview

The system records and tabulates accidents (collisions) and citations statistics in a variety of ways that reveal meaningful trends about a jurisdiction's traffic patterns. Traffic records systems can be described in terms of their inputs and outputs: inputs describe the environment and events such as accidents and citations; outputs reveal traffic patterns through tabulated reports.

The inputs and outputs of each traffic records system are unique to that system. Hence, no one description of inputs and outputs would do justice to every traffic records system. Inputs and outputs that could be found in most traffic records systems are outlined below:

Inputs

Descriptive data about the street network:

Location codes of intersections and other important street features

Traffic control devices, i.e. signals, signs, and markings and where located.

Boundaries of traffic beats

Traffic volumes on key street segments

---

1. "Societal Costs of Motor Vehicle Accidents," Preliminary Report (Washington, D.C.: U.S. Department of Transportation, National Highway Traffic Safety Administration, April 1972). Total cost to society includes the costs of accident investigation, property damage, and costs and awards of civil litigation.

Coded summaries of reported accidents (collisions):

Location

Date and time of occurrence

Type of accident, e.g. rear-end collision

Number of cars, pedestrians, and/or bicycles involved

Directions of travel by respective participants

Primary and secondary causes

Extent of property damage and bodily injury

Weather and street conditions

Coded summaries of traffic citations:

Location and direction of travel

Date and time of occurrence

Traffic code(s) violated

Reporting officer

Weather and street conditions

Disposition: guilty or acquitted

Outputs

Summaries that highlight the nature and magnitude of the traffic problem:

Number of accidents in the preceding twelve months

Numbers of accidents by type

Numbers of citations by type

List of high accident intersections

List of high citation intersections

Analytical summaries:

Lists of intersections with ten or more accidents of a specific type in the preceding twelve months

Correlations of accident causes and citations issued, e.g. locations where primary accident cause of "running the signal" occurred frequently with signal violation citations issued during the same time period

Note: The kinds of summaries produced by traffic records systems are virtually unlimited.

While a traffic records system as envisioned here would be used by the traffic supervisor to respond to the most critical traffic flow needs, the information generated by such a system may be highly useful to other municipal functions, primarily traffic engineering. Thus, it may be most feasible to develop a single system to meet both needs, and the cost of the system can likewise be shared by both budgets. At the least, the traffic supervisor can advise the traffic engineer of locations in the street network that are candidates for design improvements, if he or she has access to a good traffic records system.

Types

Many types of traffic records systems have evolved to meet the needs and tastes of different traffic supervisors. Indeed, every traffic supervisor has some sort of a system, even if it is but a blotter record of accidents and citations. Most traffic records systems encountered in law enforcement agencies today would fit into one of the classes of systems enumerated below.

Pin Map Systems

One of the simplest of systems is to simply drive a pin into a street map at the location of the accident or citation. Over a period of time, patterns will be revealed by the buildup of pins in certain areas versus the virtual absence of pins in other areas. Pin maps have the advantage of being easy to understand and use. They have several disadvantages:

Critical factors, such as time of day, day of week, type of accident or criticality, are ignored. This can be overcome by using several maps or by color coding the pin heads, but such approaches are cumbersome.

Analysis of the data, beyond what is visually evident, is virtually impossible.

Inaccuracies accumulate when pins fall out or are positioned erroneously. It is a practical impossibility to validate the accuracy of a pin map.

Manual Tabulation Systems

In these systems tallies are maintained of accidents and citations by major characteristics, such as location, type of accident, time and day of week, criticality, and number of vehicles involved. Manual tabulations are frequently used in conjunction with, and to supplement, pin map systems. Manual tabulation systems provide numerical data for analysis that are lacking in pin maps. However, analysis of data at a level of detail below tabulation categories is extremely difficult. For example, a manual tabulation system tallies accidents both by type and by location. But the number of rear-end collisions at a specific intersection, for example, may end up in the aggregates and not as a separate figure.

Automated Systems

More and more jurisdictions are adopting these systems, particularly with the financial support of the National Highway Administration and the Law Enforcement Assistance Administration. In these systems the basic characteristics of accidents and citations, together with other relevant data to describe the street network and traffic beats, are fed into a computer operation where the data are organized and rapidly disseminated as computer print-outs. The print-outs may be available routinely in weekly reports, or they may be requested as needed for specific purposes. As an example, detailed summaries of accidents involving one or more drivers "under the influence" could be requested late in the year to help plan a holiday season campaign to detect drunken drivers and remove them from the streets.

The principal advantage of an automated traffic records system lies in its capability to rapidly and accurately sift through large volumes of data and then prepare meaningful summaries which the traffic supervisor can utilize to improve traffic enforcement and control. The principal disadvantage is the cost of developing and installing an automated system. On balance, though, automated systems are becoming ever more prevalent, and the student can reasonably expect to be involved with one in the not too distant future.

Uses

To be effective, the traffic function must plan, execute, and control the various actions necessary to accomplish intended objectives. A traffic records system is concerned primarily with the first and third of these: planning and controlling.

A vast amount of data is gathered and processed by a traffic records system to reveal patterns of traffic activity. Revelation of such patterns will lead to ideas about actions needed to effect improvements. These ideas are then formulated in specific plans for the future. In documenting these plans the traffic records system provides additional data regarding the availability (or lack) of resources to execute the plans. The system can also indicate the successes and failures of previous plans with similarities of scope and/or intent.

When plans are executed, the planner needs informational feedback to determine if the plans are working or if corrective actions are needed. A traffic records system provides the required feedback for evaluation of the plans and to formulate needed corrective actions.

Several specific uses of traffic records systems for planning and controlling are listed below:

Establish the magnitude and nature of the traffic problem. As a matter of policy, the chief law enforcement officer and other officials need information as to the total number of accidents and violations, the impact of these on life and property, and whether the year to year trend is improving or getting worse. This information is essential for budgetary planning and justifying the adoption or continuation of traffic improvement programs. At a minimum, this information should include:

Total accidents of all types, and summarized by injury/noninjury, property damage, whether pedestrians were involved, and by major cause.

Total citations issued and summarized by infraction. The summaries should also indicate what fractions of the citations resulted in convictions.

Provide surveillance of high accident locations. Certain intersections and roadway segments are more prone to accidents than others. The traffic records system should identify these locations and provide data to indicate whether or not plans to reduce accidents are working. A good traffic records system will have the capability of preparing a detailed summary of all accidents (within a specified time period) at a given location. A study of these summaries can provide clues as to the underlying traffic flow problems, and traffic enforcement plans can be upgraded accordingly. For instance, if a given intersection's accident pattern is dominated by rear-end collisions in the westbound side, then the traffic supervisor should issue orders to step up citations for following too close in the appropriate street segment. He or she could also recommend to the traffic engineer that a "caution" sign or a "stop ahead" sign should be posted to warn drivers of a stop ahead.

Permit selective enforcement by location and type violations. Because a traffic records system provides detailed knowledge of "what's happening where," the traffic unit can be deployed so as to concentrate on select areas of need. The traffic unit should be alerted to specific violations which should be enforced.

Balance traffic workload according to need. The data provided by a traffic records system can be an important input to the process of establishing beat boundaries and to justifying the separation of traffic enforcement from general patrol in certain areas. The lack of such balancing can lead to the aggravation of traffic problems.

Determine effectiveness of traffic safety programs. From time to time new programs will be adopted which aim at improving specific aspects of traffic, for example, a program aimed at lowering accidents involving youthful drivers. The traffic records system can be an invaluable aid in measuring the effectiveness of such a program. In our example here, the traffic records system could prepare a comparative report of accidents by "Driver Age" for that program period and the same period in the prior year. If an improvement is *not* evident, it is obvious that adjustments to the program are in order. And if an improvement *is* evident, then the traffic supervisor is in a stronger position to get his or her next improvement program approved by higher authorities.

Identify needs for driver/pedestrian education programs. The traffic records system can point out various needs for education. For example, if accidents involving pedestrians should increase around a particular commercial district, the traffic supervisor could initiate plans for a display of posters in that district to make both drivers and pedestrians more conscious of pedestrian safety needs.

Provide data for officer evaluation. A traffic records system can summarize enforcement activity by the initiating officer. The traffic supervisor can then determine which officers are responding to orders regarding selective enforcement, and which are not. Furthermore, a traffic records system can discourage the abusive practices of operating a "cherry patch," i.e. in issuing numerous citations for insignificant violations, such as exceeding the speed limit on a stretch of road which has little traffic and no hazards. ("Cherry patch" alludes to issuing numerous citations with little effort, analogous to picking cherries in an unpicked orchard.) In a more positive vein a traffic records system can be the basis of incentive programs and for recognition of outstanding officer achievement in implementing traffic programs.

Problems

Nearly every traffic enforcement agency has some form of a traffic records system, and no attempt will be made here to survey *all* of the problems of all traffic records systems. However, the proliferation of automated systems is bringing a distinct class of problems to the using agencies, and they will be touched on here.

The biggest problems associated with an automated traffic records system involve its installation. Because development costs are typically high, an early problem is to get financial support of the development budget. Assistance from the federal government in the form of a grant is most helpful in this regard.

Once a budget is established and go-ahead is authorized, the actual development can begin. The development involves an intricate design effort by computer specialists. Several problems can arise in such an effort: poor planning, lack of clear communication between police and technicians, conflicting objectives and priorities, changes in objectives, policies, and withdrawal of financial support, to name a few. This can be softened somewhat by copying an existing traffic records system that is operational in another jurisdiction, but even an existing system must be tailored to local needs.

As the development nears completion, the problem of getting people to alter existing work patterns emerges. The new system generally substitutes new report formats for old ones and introduces reports that were never before available. The traffic analyst finds the comfort of his or her established routines badly shaken by this change. He or she must be encouraged to bear with the situation until his or her professional equilibrium can be regained. The traffic supervisor should not allow the analyst (or for that matter himself or herself) to sabotage the effort because of its strangeness. The supervisor should stress the benefits of the new system to all persons involved so as to enlist their enthusiasm in making the change.

Traffic records systems have in common with all other information systems the persistent threat of GIGO. GIGO is an acronym for "garbage in, garbage out." It simply means that if incorrect or incomplete data is entered into the system, the summarizations will be accordingly incorrect. Frequently, errors are magnified many times, e.g. an incorrect date may show up many times. Quality control of data handling is extremely important because there are so many opportunities for mistakes to be made: the officer's illegible handwriting on the source document; omission of accident/citation reports or duplicate reporting; mechanical mistakes by machine operators; and subtle computer program errors that may not become evident for years. While much of the data handling may be beyond the control of the supervisor, he or she can set a good example for that portion of data handling that is his or her responsibility, and he or she must insist on quality performance by the agency that processes his or her data. Frequent audits and spot checks of output data by the supervisor and staff are most helpful in this regard.

A traffic records system will produce a voluminous amount of information in the form of tabulated summaries. A generous amount of time must be allocated for studying these summaries in detail. While these summaries convey traffic patterns not discernible otherwise, such patterns are seldom obvious from a cursory glance. The supervisor and his or her key staff subordinates must immerse themselves in the data to discover the meanings therein. For example, a particular intersection might have relatively few accidents. But if nearly all of them are collisions where one driver was attempting a left turn, a need for some kind of traffic improvement is evident.

Finally, the supervisor on discovering many opportunities to improve traffic control with his or her traffic records system must resolve to follow through with plans and programs to do something about his or her findings. Doing so, his or her greatest reward may be the sense of satisfaction he or she discovers, through the traffic records system, that his or her plans and programs are working!

Future Directions

A continuing thread in this chapter has been the appearance of automated traffic records systems in great numbers. That trend will surely continue as computing equipment becomes cheaper and more readily available.

Some jurisdictions have found it advantageous to band together with regional systems. A number of cities and counties near San Francisco, California, are participating in a regional system with very advanced computer and data base management techniques.

Other regional systems have been developed, or are under way, in Los Angeles and Orange Counties in California and Oakland County, Michigan. Regional systems offer several advantages: cost-sharing, the ability to use high-powered equipment and techniques, and the capability for each jurisdiction to compare its own performance against regional norms. The principal disadvantage is securing cooperation of all agencies in all matters. Decisions as to who shall exercise overall authority and control are especially vexing.

The development of super-fast computers with large memories at low cost is opening up the possibility of interactive traffic records systems. In such a system the analyst will be able to work at a console and manipulate the system. If he or she discerns a pattern of "failure to yield" violations at a particular intersection, for example, he or she could immediately order a display of the details of each such accident. In current systems he or she would have to order additional summaries and wait several days to receive them. In the meantime, his or her train of thought could be lost.

If the traffic records system utilizes a computer system that is shared with other governmental functions, the traffic records may be integrated with other data bases. For example, suppose the Planning Department maintains small area population statistics furnished by the U.S. Bureau of the Census. The commander may discover that accidents involving older pedestrians are more predominant in areas of the community with high median ages. He or she could then initiate a program to afford better protection to these senior citizens.

Summary

A traffic records system was defined as the collection of documents, files, procedures, and summaries that provide a history of traffic problems in a jurisdiction in a way that permits assessment of overall cause and effect. It is far broader than report preparation and filing procedures. Traffic records systems were examined from a supervisor's perspective as management tools that can improve upon the overall traffic problem.

The traffic supervisor's purpose in having a traffic records system is to obtain informational feedback about his or her plans and actions to improve traffic control. Thus, a traffic records system's major benefit is for planning activities.

Traffic records systems record and tabulate accident and citations (summons) statistics so as to reveal meaningful trends about a jurisdiction's traffic patterns. Systems can be described in terms of their inputs and outputs. Inputs include descriptive data about the street network, coded summaries of reported accidents (collisions), and coded summaries of citations. Outputs include summaries that highlight the nature and magnitude of the traffic problem, and analytical summaries to reveal particular causes of general problems.

Traffic records systems are also useful to other municipal functions, primarily traffic engineering. Cooperative development and cost sharing of a single traffic records system to meet both needs is often feasible.

The most common classes of traffic records systems that the student might encounter are: (1) pin map systems, where accidents and citations are recorded with pins stuck in a map; (2) manual tabulation systems, where hand tallies are made of types of accidents and citations; and (3) automated sytems, where all input data is fed into a computer system for rapid processing and dissemination. Of these, the automated systems are generally more effective and are becoming more prevalent.

A traffic records system is used primarily for planning and controlling actions to achieve traffic investigation and control objectives. Some specific uses are:

Establish the magnitude and nature of the traffic problem

Provide surveillance of high accident locations

Permit selective enforcement by location and type violation

Balance traffic workload according to need

Determine effectiveness of traffic safety programs

Provide data for officer evaluation

A number of problems are associated with establishing a traffic records system, especially if it is to be automated. The biggest problem is getting it installed—finding financial support and mastering the technical development. Another problem is that of getting people to accept the changes incurred by a traffic records system. Still another problem is that of quality control so that summaries are both timely and accurate. The supervisor must face up to the challenge of re-ordering his or her own schedule so that he or she devotes a generous amount of time to studying the voluminous amount of data given out by the traffic records system. In doing so, his or her greatest reward may be the satisfaction received when he or she discovers from his or her traffic records system, and because of it, that his or her plans and programs are improving traffic safety and control in his or her jurisdiction.

In the future, more traffic records systems may be developed on a regional basis, such as the system being used by communities near San Francisco, California. Future improvements in computer technology may result in interactive systems where the analyst develops his or her information in a dialogue with the computer. Finally, the integration of the traffic records system into other municipal information systems is possible where data bases can be shared.

DISCUSSION QUESTIONS

1. A summary from your traffic records system shows that a particular intersection has had twenty accidents in the last year in which the primary cause was "running the signal." Would you talk to the traffic engineer about improving signal visibility? Would you issue special orders to the cognizant beat officer? What other actions would you consider?

2. Your year-end traffic summary indicates the following:

	This year	*Last year*
Total accidents	1,382	1,394
Fatal accidents	20	22
Multiple car accidents	330	398
Single car accidents	1,052	996
Accidents involving pedestrians	68	49
Accidents involving bicycles	53	50

What is your interpretation of these figures?

True—False

1. In a study of traffic records systems, it is most important to become intimately familiar with the procedures for writing field reports and how they are indexed and filed by the Records Bureau.

2. The prevention of accidents from information revealed by a traffic records system can be thought of as "worth" $2,800 per prevented accident to society.

3. An important input to a traffic records system is descriptive data about the street network.

4. Analytical summaries are a luxury that should not be expected as an output of a traffic records system.

5. Pin map and manual tabulation systems are tending to displace automated systems.

6. A traffic records system cannot offer any relief to the problem of balancing workload.

7. A traffic records system improves traffic officer efficiency by encouraging the development of "cherry patches."

8. An early problem of installing a traffic records system is to get financial support of the development budget.

9. Typical problems encountered in developing an automated system are the lack of clear communications between police and technicians, changes in objectives, and politics.

10. Unless he or she perseveres with quality control standards, the supervisor is likely to get "garbage" from his or her traffic records system.

5: Enforcement Tactics

Public support for the enforcement of traffic violations is of utmost concern to law enforcement agencies today. The success of the police in reducing traffic accidents hinges on the methods employed in traffic enforcement. The police have the responsibility to move traffic in the most expiditious yet safest manner possible. The accidents that are largely created by the thoughtless motorists who violate the law by speeding, cutting in and out of traffic, following too closely (tail-gating), and so on will always continue to be a problem. Although there are presently many types of enforcement programs being employed throughout the country, only a few of them are having a deterrent effect. One of the most effective methods to reduce these high accident exposure areas is the use of selective traffic enforcement. This is a somewhat negative tactic as far as the public is concerned, because it employs maximum enforcement in a concentrated location. The purpose is to take enforcement action on those traffic violations that are causing or tending to cause traffic accidents. The public, however, feels that the police are being overzealous in their enforcement. This has had some detrimental effect on the public's support of its local police agencies. It should be pointed out that it is really not the tactics generally utilized by the police that bring the charges of "extremism," but rather the lack of "public information" by the police as to the reasons for the use of such tactics.

The need to "sell" highway safety to the public is a continual, ongoing challenge to American law enforcement agencies. The use of various highway safety programs and the education derived by the public from them will in turn increase public support while at the same time bring about accident reduction. To this end we shall explore the problems, possible alternatives, and hopefully the rationale for the use of sound traffic enforcement tactics within this chapter. We will not present the one best way to deploy personnel and equipment to reduce accident severity, but rather a number of considerations useful for the methodology of traffic enforcement.

First let us consider the two individuals that we will focus on in order to obtain a clear understanding of the actual problem—the traffic violator and the traffic enforcement officer. The driver (and all of us fit into this catagory at one time or another) of any motor vehicle has both a legal and moral responsibility to drive in a safe, reasonable manner. This of course must also include all law enforcement personnel assigned to mobile enforcement and/or patrol. Drivers may be placed into one of three main classes: (1) the defensive driver, who seldom commits a traffic offense; (2) the occasional offender; and (3) the chronic offender.[1] It is the last two classifications that our second individual, the police officer (hereafter referred to as the "officer"), will direct most of his or her attention. While we realize that a large number of drivers fall into the classification of the "occasional offender," there are still far too many drivers falling into the "chronic offender" classification. Both warrant selective discussion. The occasional offender usually, for some thoughtless moment, takes a chance at beating the traffic light that is just about to change from caution to red or stop. Perhaps the individual may be late for work or school and so speeds up a bit, in order to reduce the amount of time it will take to get from home to his or her destination. He or she does not normally drive this way, but circumstances have caused the driver to modify his or her driving behavior to meet the problem of tardiness. The chronic offender is a socially maladjusted individual who regards our system of social order, which is a series of regulated legal statutes, as being unworthy of his or her attention. The chronic offender feels the law was always made for the other person, and not to provide guidelines for an orderly and civil society. This driver, although having received numerous traffic summons or citations (depending upon the particular part of the country), continues to violate the law, thus becoming a mobile threat to the safety of the motoring public.

1. Paul B. Weston, *The Police Traffic Control Function* (Springfield, Ill.: Charles C. Thomas, 1969), p. 41.

The officer's job is to protect society from the drivers just identified. In reality, the officer finds himself or herself in a quandry trying to establish some type of human balance between enforcing the law by the spirit rather than by the letter of the law. This man or woman is highly trained, works under rules and departmental guidelines, obeys the law, and has a code of ethics. It is this individual to whom the major emphasis of this chapter is geared.

Objectives of Traffic Patrol

Although there has been little written on some of the very basic police methods and techniques otherwise referred to as "know-how," experience has provided some guidelines. The patrol officer assigned to a line or area beat learns continually from older, more experienced, as well as fellow, officers. The advice that he or she receives and the skills that he or she develops are the result of observation and analysis.

The learning ability of officers, and, for that matter, people in general, may be divided into three types: (1) those who learn by others' experiences; (2) those who learn by their own experiences; and (3) those who never learn.[2]

There is a great deal of statistical evidence to validate the fact that many officers have had a tendency or have in fact fallen into the second and third groups. The everyday business of issuing traffic summons or citations, as well as making physical arrests, provides the best type of field instruction for other officers. There is really no one best way to perform these duties, as circumstances will vary from jurisdiction to jurisdiction and from state to state. Some guidelines may, however, be provided to assist the officer in safely and efficiently carrying out these tasks.

First of all, one must be mentally prepared. The officer must be in the right frame of reference, that is, "knowing that I am a police officer—I have a duty and a responsibility to the public for which I serve." Enforcement is not just a job assigned to an individual, but a professional undertaking that involves judgment, skill, and a high degree of training.

Second, the officer must be provided with the necessary tools with which to work. These include but are not limited to: state traffic or vehicle code statutes, departmental guidelines for proper enforcement, technical and legal training, and authority and responsibility within which to function. In addition the

2. *Enforcement Tactics Manual* (Sacramento: California Highway Patrol, 1965), p. 1.

officer needs the hardware, i.e. patrol cars, cruisers, motorcycles, motor scooters, helicopters, fixed wing aircraft, radar units, and so on from which he or she must operate. These tools are the basis by which the more technical methodology is developed.

The officer through observation mentally checks the actions of his or her fellow officer and determines whether a certain procedure employed is good or bad. He or she then has the responsibility to incorporate, as a positive trait, that which is good and reject that which he or she finds to be bad.

A symbolic purpose of American law enforcement is reflected in this quotation from Abraham Lincoln: "The first intent of our American government is not to compel people to do certain things and restrain them from others, but rather to make the right life and the useful life the natural and easy one to live." Officers should strive to mirror this example and support its ideal. The officer is the voice of authority and not the authority, for that rests with the people. [3]

The last main ingredient to the formation of good sound traffic enforcement is the support of the motoring public. This does not necessarily mean that the general public should enjoy the mass issuing of traffic summons or citations, rather it means that the public understands the need for highly professional traffic enforcement and helps by complying to a greater degree with the various traffic statutes of each state. Although the basic role of police and traffic patrol is to reduce the opportunity to commit unlawful acts, it is encumbent upon the general public to follow the rules and guidelines of highway safety.

The intelligent officer, through careful patrol methods, is the first line of defense against accidents. The omnipresence, or the impression of it, is created by frequent and conspicuous patrol around the clock. Adequate patrol succeeds in effecting immediate apprehensions and building a reputation for quick, courteous, and professional enforcement which is passed on through the media and from person to person.

There are other duties that the enforcement officer must perform while on routine patrol. For instance, there are numerous types of public services: removal of traffic hazards such as rocks, broken glass, tree limbs, spilled loads, animals on the roadway; directions, travel information, and other similar types of aid; assistance to disabled motorists; and checking and removal of abandoned and stolen vehicles.

---

3. *Enforcement Tactics Manual*, p. 1.

The public obtains a picture of the type of law enforcement agency present within the community based on what it sees in the daily activities of the patrol officer. It is the basic patrol function which occupies so much of the individual officer's time; therefore, he or she must employ both efficient and effective enforcement tactics.

By patrolling, officers are better able to provide better services, but they also are in a position to give greater protection to the driver. There are four primary purposes for patrolling: observation, apprehension, deterrent effect, and assistance to the public. Also through setting an example officers encourage safe driving practices. Officers should always drive or ride in their cruisers or patrol vehicles in such a manner as to present the impression that they are "on patrol." Being alert and available while on a tour of duty is an officer's continued responsibility. By doing this the officer indirectly instills in every driver six fundamental beliefs.

1. It is the driver's responsibility to know and understand the law.
2. He or she must believe in the law.
3. He or she must have the feeling that the traffic officer is everywhere.
4. If the driver violates the law, he or she will be caught.
5. When a driver is apprehended he or she will be arrested.
6. Finally, upon arrest and conviction or adjudication there will be punishment.

Techniques of Traffic Patrol

Visible Enforcement

One of the most effective methods of reducing the opportunity for suspected wrongdoing is through visible traffic enforcement. The potential violator is deterred from committing a hazardous violation by the very fact that a uniformed officer, in a distinctively marked patrol unit, is moving or at rest upon the highway. An open, active, and visible patrol affords better supervision of traffic and, since a greater area is covered, a deterrent effect results.[4] There is no deterrent effect when an officer places himself or herself and his or her unit in an area where he or she is virtually in

4. *Enforcement Tactics Manual*, p. 9.

hiding. There are some cases, however, where the traffic problem is so great and the hazard is of such a degree that it may warrant an officer to be in a position of advantage in order to observe those accident-causing violations. The fact still remains that where there is a high degree of officer visibility and the public is very conscious of his or her presence, there is a definite reduction in accident-causing violations.

What is the traffic enforcement officer actually doing while on routine patrol? He or she is in a position of readiness, waiting for a call (a request for service from the public), and observing (looking for those conditions that may require service or enforcement). The former is reactive in that the call for his or her services stems from a source to which he or she must respond. The latter is proactive because the patrol officer is seeking to initiate remedial action in order to solve a particular problem. The major function of a patrol officer is observation. Clearly, the officer's effectiveness is dependent upon that which he or she observes and evaluates. The scope and sureness of his or her observation has an influence upon his or her ability to detect problems of varying degrees. He or she is able to detect offenders, remove hazards and provide for general services.[5]

The officer must plan his or her daily tour of duty. This means that he or she carefully evaluates the problems that have occurred from the time he or she went off duty until his or her return. He or she must read the daily log, teletypes, special broadcast memorandums, departmental directives and orders, and so on. During the role call or briefing session the officer should pay particular attention to those occurrences that will have some effect upon his or her patrol operation. Those things that his or her supervisors identify as being of utmost importance should be noted in written form in a pocket notebook for future reference. Just before going out on patrol, the officer should check with the beat officer just coming off duty who had been assigned to the same area of responsibility. The officer should determine any unusual problems or hazards that may not have been passed on to him or her through the briefing session. It is always wise to get a general feeling of how the activity has been on the previous watch or shift. If there has been an extremely busy time just handling traffic accidents or "crashes" (a new term that is replacing the words "collision" or "accident"), then the officer should plan to cover this area first upon going out on patrol. Finding out what

5. Paul M. Whisenand, James L. Cline, *Patrol Operations* (Englewood Cliffs, N.J.: Prentice-Hall Essentials of Law Enforcement Series, 1971), p. 10.

type of traffic violations have been causing these accidents may warrant him or her to perform some diligent type of enforcement in these areas.

Although some departments sanction the use of off-street parking for observation while others do not, there is a need for this activity at times. It is possible that there is no other area immediately available or the potential hazard to the officer may be too great to maintain a fixed post upon the surface street. A check with the department's policy and/or the supervisor should determine if this is an allowable method of fixed observation. If the method is used, there are some cautions to keep in mind. It is important that the officer always park the patrol vehicle legally. The patrol unit is never parked on the wrong side of the street, headed in the wrong direction. This creates a hazard to both the officer and to other drivers in the area. Citizens may feel that the officer is attempting to hide and, therefore, it becomes a kind of game with them. Drivers also resent deliberate violations of the law by patrol officers. This tarnishes the very image of professionalism that we are trying to create.

Officers should avoid becoming too set in their patrol procedures. By varying their activity through careful planning they become more alert and eager to perform their daily tasks. The very reason for their presence is to reduce the potential crash. The officer who becomes so methodical in his or her day-to-day patrolling procedures diminishes his or her effectiveness proportionately. It is always a good practice to reverse the patrol, for example, by stopping the vehicle at times in conspicuous locations, holding back from the entrance to an intersection at a red signal to observe cross traffic, circling the block and approaching the intersection from a different location, and operating in the extreme right-hand or outside traffic lane, so as to be readily available when the need arises. By changing the pattern of patrol, other drivers cannot predict an officer's presence and, therefore, the effect of omnipresence is established (the officer appears to be everywhere). It may be that someone is waiting for the officer to pass by in order to commit some unlawful act (other than a traffic offense). Another effective technique is to slowly approach an area of concern, then pull up and stop at a location where the officer can observe carefully without being seen too easily and wait for a few minutes. This may fool even the most cautious deliberate violator or criminal. Remember the deliberate violator likes nothing better than to know where the patrol unit is. By maintaining a methodical type of patrol, drivers soon learn to anticipate an officer's presence at a given location and will regulate their conduct accordingly. This, however, may be a positive factor at times. For example, the problem of drivers passing a stopped school bus with its flashing red lights operating and children exiting or entering the school bus is deadly. When this comes to an officer's attention, he or she should take positive and immediate steps to deter it. Placing himself or herself at the various locations along the school bus route, during the time of its operation, and apprehending those violators that pass the stopped school bus may greatly curtail the hazard. If an officer is observed in the area routinely enough, that in itself may all but completely eliminate the problem.

Speed of the Patrol Unit

The proper speed of an officer's patrol vehicle has a great deal to do with the quality and quantity of his or her apprehensions. While the speed of the officer's patrol vehicle may vary with the type of enforcement being used, or the location (city streets versus expressways and freeways) and the traffic with which he or she must contend, there are some fundamental principles to be considered.

Failure of the officer to observe traffic violators while driving at high speeds is not uncommon. This is due in part to the fact that at higher speeds one's vision periphery tends to narrow to the point of tunnel vision. In fact tests have proven that at speeds in excess of 80 to 90 miles per hour one is no longer steering a vehicle but aiming in excess of 4,000 pounds of hurtling steel. The chances are far less at high speeds of being able to negotiate a vehicle should a sudden emergency arise. Further, high speeds appear to the public as an abuse of an officer's immunity to speed regulations. Although each department's recommendations for normal patrol speed may differ slightly, it is suggested that a speed of five miles less than the normal flow of traffic be maintained. This allows the officer the greatest opportunity to observe other drivers, while at the same time operating his or her patrol vehicle at a speed which will afford the best maneuverability should a violation be observed. It must also be pointed out that driving at a speed too slow may impede the flow of traffic, thereby creating a hazard as well as possibly violating the law. The patrol unit should always be driven in the extreme right-hand lane during routine patrol duties. However, when using emergency lights and siren, the unit must be driven in the far left-hand or inside traffic lane. This is in order to give the emergency vehicle the safest portion of the highway opon which to operate. When operating on narrow or two-lane roadways the officer should frequently pull over to the righthand curb or edge of the roadway, and in some cases, actually stop momentarily to allow traffic to pass.

Apprehending the Traffic Offender

Once a traffic violation is observed, the officer must be in a position to effect an arrest thereby taking some type of positive enforcement. The officer has a responsibility to apprehend the offender as quickly as possible. This reduces the opportunity to evade arrest or to create additional exposure to other drivers who could very well be involved in a traffic mishap with the violator and prevents the offender from forgetting about the traffic violation and possibly developing some type of alibi. It also affects other motorists, who may have witnessed the violation, by assuring them that violations of the law will be dealt with swiftly and positively. This instills in other drivers that law enforcement is for the protection and safety of all upon our streets and highways. This in turn generates public support for all law enforcement.

Before discussing the techniques of apprehension, some background factors should be covered. The first and most important consideration is the officer. Has he or she been properly trained in the operation of an emergency vehicle? This should be a combined classroom (about 20 percent) and actual driver training (about 80 per cent) type of emergency vehicle driving course (see figs. 5-1 and 5-2). Most states having large police agencies have their own Emergency Vehicle Operations Course (E.V.O.C.) such as those of the California Highway Patrol, Michigan State Police, Ohio Highway Patrol, New York State Police, Los Angeles Police Department, and the Los Angeles County Sheriff's Department. There are also privately owned and operated defensive driving academies. These are usually instructed by former police officers and race drivers, all well qualified in the skills of defensive and emergency driving techniques. Most of these private organizations are in some way associated with various law enforcement agencies in the basic driver training of new police officers. One such organization is the Academy of Defensive Driving in Irvine, California. This school has been set up by retired and inservice law enforcement personnel, all of whom are experts in the field of emergency vehicle operations and defensive driving techniques. For example, this school meets the high standards for police driver training as required under the Peace Officer Standards and Training (P.O.S.T.) in the state of California. There are a number of such schools, both public and private, to which a police department can and should send each officer for training.

FIGURE 5-1
Classroom Instruction

FIGURE 5-2
Field Training and Testing

The minimum recommended course should be no less than three days (twenty-four hours) of actual contact time.

The officer should be familiar with the provisions of his or her respective state's vehicle code, to determine what the limitations are of criminal and civil liability as they pertain to the operation of an emergency vehicle. An officer must also be well acquainted with his or her individual department's orders, policies, and standard operating procedures as they may pertain to the operation of a police emergency vehicle. We must point out that governmental agencies are no longer exempt from civil liability and responsibility resulting from the negligent operation of any emergency vehicle while going code three (operating with visible, distinctively colored emergency warning lights, and an audible, operating siren).

The second consideration is the type of high-powered police vehicle being used. Most law enforcement agencies provide specially equipped vehicles for use in apprehending law violators. Each officer must respect and operate these vehicles as if they were his or her private vehicle. This is not to say that when the need arises he or she should not effectively use the power provided to rapidly overtake an offender, but rather operate the police unit within the realm of good sound judgment. It is very important that the officer make a complete inspection of his or her police patrol car before going out on duty. This includes both the exterior and interior: all lighting equipment, both regular and emergency, the siren, tires, brakes and steering, and any special emergency, equipment that may be provided for his or her use in addition to that which is a permanent part of the police car. It is his or her duty and responsibility to report any malfunctions, damage, or missing items from the patrol car before turning it over to the next officer for his or her tour of duty.

Now that we have discussed the two considerations significant to the successful operation of a police emergency vehicle, let us look at the particular techniques for apprehending a violator.

The officer must be well acquainted with his or her beat and should know those areas where a violator can be stopped safely and clear of traffic. He or she should select the proper place to stop the violator, keeping in mind both his or her safety and that of the violating driver. At night it is best to make a traffic stop within a well-lighted area whenever possible. This practice will materially reduce the chances that the occupants of the violator's car will resist the officer or endeavor to overpower him or her. Likewise, there may be other persons in the area who will serve as witnesses to what transpired during the physical arrest or issuance of a citation.

After having selected the location along the roadway at which the officer desires to stop the violator, he or she should check other traffic around him or her, particularly that to the rear. Use of rear emergency warning lights, to advise drivers following the patrol unit that a traffic stop is about to take place, should be employed at least until the hazard no longer exists. (This of course, may vary by department policy and the needs of the officer—his or her safety—at that moment.)

The officer uses emergency warning lights to the front and sounds the horn or siren (if needed) to gain the attention of a violator. As soon as the violator responds to these signals, the officer starts braking the patrol vehicle, bringing it to a stop between ten and fifteen feet to the rear of the violator (see fig. 5-3 and 5-4).

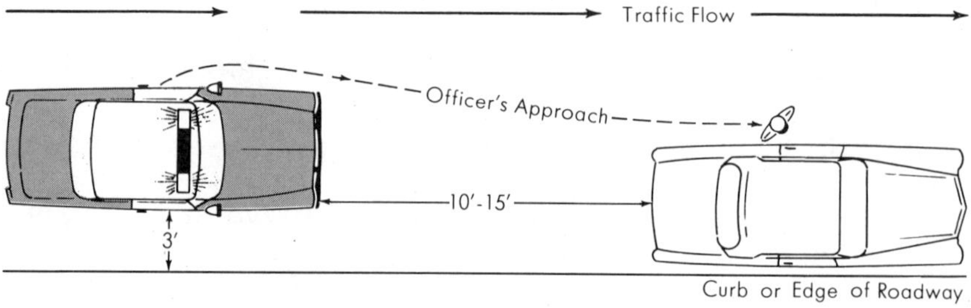

FIGURE 5-3
Left Off-Set and Left Approach

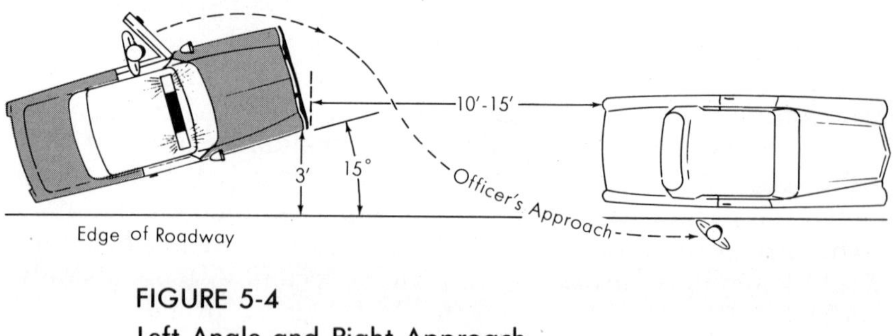

FIGURE 5-4
Left Angle and Right Approach

The single officer approach to the violator's vehicle may be either from the left or right side. If made from the left side the officer should always check the condition of traffic approaching from the rear before exiting on the driver's side. Many officers have been seriously injured and some even killed while alighting from their patrol cars on the left side. On the other hand, the right-hand approach by the single officer has proven a very effective method of making an enforcement contact. It gives the officer that extra margin of protection from the oncoming traffic, the element of suprise, a better point of observation, and a somewhat more accessible place from which to contact the violating driver. In figure 5-3 a single officer approach is shown, left off-set and to the rear of the violator. Figure 5-4 is similar but in this case the police car is parked at a 15-degree angle and the officer now makes a right-hand approach rather than a left-hand approach.

Two other methods of single officer traffic or other type of enforcement stop are shown in figures 5-5 and 5-6. In figure 5-5, the traffic stop is made in the conventional method, probably on a city street or other similar location. The police car is parked directly to the rear of the violator's vehicle, leaving a safety space of between ten and fifteen feet. The officer may elect to make his or her approach from the left side or from the right side as is shown in the diagram. The other method, shown in figure 5-6, is used primarily where there is a very narrow access. Note that the police car parks off the road slightly. The left side of the police car is almost in line with the right side of the violator's vehicle. At this point the officer may use either the left or right side approaches.

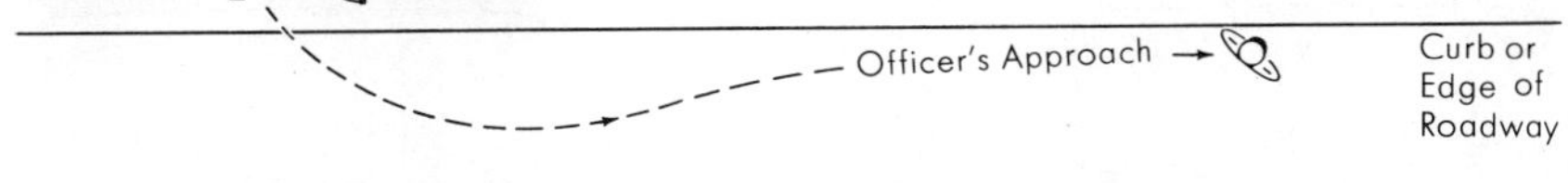

FIGURE 5-5
Parking Directly to Rear—Right Side Approach

When there are two officers working together in the same police car, the method of stopping the traffic violator and the approach may vary from department to department. The following two diagrams illustrate methods for making high-risk and/or felony stops. Figure 5-7 shows the police car parked straight away or directly to the rear of the violator's vehicle with a safety space of

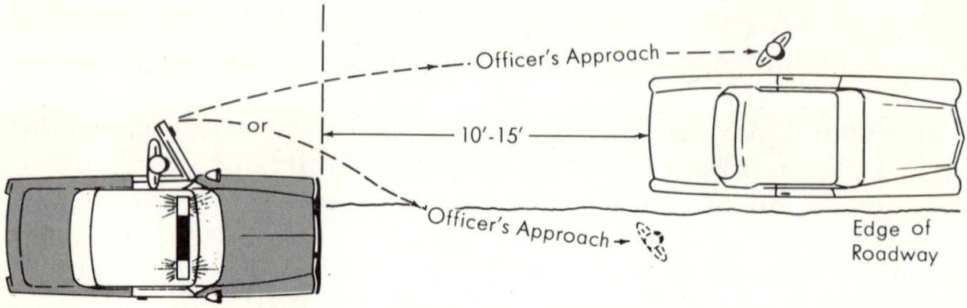

FIGURE 5-6
Narrow Access—Right or Left Approach

between ten and fifteen feet. As the driver of the police car makes his or her left-hand approach, the passenger-officer alights from the right side of the police car, hesitating momentarily behind the right passenger's car door. After the officer making the left-side approach makes contact with the driver of the violating vehicle, the passenger-officer moves to a position at the right front fender of the police car. The passenger-officer should stand about one foot to the right of the right fender so that he or she may be in the best possible position of observation. It may be necessary in some cases for the passenger-officer to move up to a position along the right-hand side of the violator's vehicle. Caution should be used in this circumstance as it leaves both officers without any protection. Should an emergency occur the passenger-officer has considerable distance to travel in order to return to the police unit. This last method is illustrated in figure 5-8, using the left off-set technique of parking the police car.

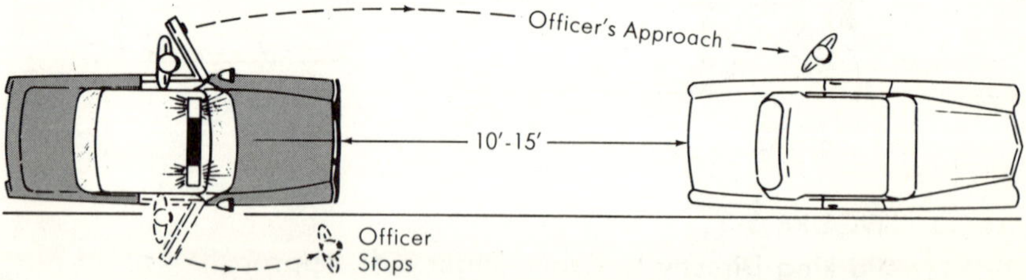

FIGURE 5-7
Straight Away—Two Officer Patrol Unit

Upon making verbal contact with the violator, the officer should explain the purpose of the stop and request that the driver's license and vehicle registration be presented. He or she should briefly discuss the conditions surrounding the violation

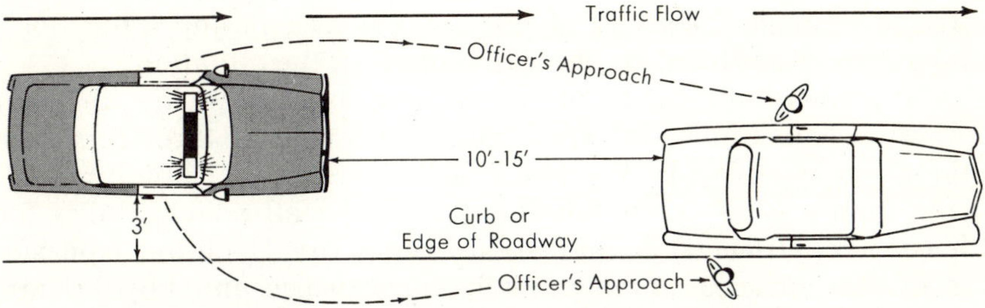

FIGURE 5-8
Approach of Two Officers with Left Off-Set

and explain the type of enforcement about to be employed (verbal warning, written warning or citation, or physical arrest). The officer should then terminate the contact in a professional manner, allowing for any questions from the driver. The officer also should explain how the driver is to take care of the summons or citation, by appearance or by mail, and so on. Remember, the officer is not only enforcing the law but also selling highway safety. He or she should allow the driver to merge with traffic safely before proceeding on. The officer should also remember to always verify the driver with his or her license and the registration or title with the vehicle he or she is operating. Many automobiles are stolen each day of the year and part of an officer's sworn duty is the apprehension of these criminals.

The officer often uses other forms of traffic regulation to accomplish his or her purpose. One is through hand traffic direction (see chapter 7 on traffic direction). Another is through the promotion of pedestrian, bicycle, and motor vehicle safety education programs. Neither time nor space provides the authors an opportunity to explore and discuss these programs, so it is suggested that the officer investigate the types of highway safety programs that are ongoing in his or her community. Class discussion may generate the numerous pros and cons of these worthwhile programs and perhaps at the same time establish some direct needs in this area.

High Speed Pursuit Driving

Pursuit and emergency driving involve the same driving skills that the field officer must master for normal routine patrol operation. There is the additional emphasis on good sound defensive driving skills. Pursuit driving requires that the officer be at peak efficiency and extremely alert, a condition that is actually enhanced by the increased activity of the nervous system and other body functions, stimulated by the excitement of the situation.[6]

Of prime importance to the officer in making the decision whether to go into a high-speed pursuit, or a "code three run," are the circumstances. If the officer feels justified in going into pursuit, then it is he or she who must shoulder all responsibility for his or her actions while on the emergency run. Deciding to pursue is always a process of weighing the hazards presented by the pursuit as opposed to those hazards created by allowing the violator to escape arrest. Although there are varying degrees attached to each circumstance, it is the officer who must rely on his or her good judgment to make the final decision. The degree of hazard in which the officer must place himself or herself and others is dependent upon the degree of hazard in which the violator threatens other traffic.[7]

High speed chases often begin with a simple traffic violation. The officer, while attempting to stop the violating driver, finds to his or her amazement that not only is the driver not stopping, but also he or she is actually trying to evade a lawful arrest. While in pursuit it is always a good idea to reevaluate the situation to determine if it may be wiser to obtain more information about the fleeing suspect and the automobile he or she is driving, to follow up with an investigation, and then to file a complaint in the local court. Officers have lost their lives while in pursuit of a traffic violator who initially ran a red light or a stop sign.

Experience has shown that use of the siren and red light does not always assure an officer of immediate right-of-way. Conflict occurs many times because a motorist is unable to hear the siren. The human element is always present and under emergency conditions motorists and pedestrians will react differently. Certain techniques in the use of the siren and in driving will compensate to some extent for the shortcomings of persons who are excited or inattentive to their driving.

The siren should be operated throughout its entire tone scale, fluctuating from a high to a low pitch. Where the possibility of traffic conflict is present, the siren should be activated sufficiently in advance so that pedestrians and motorists may have adequate warning of the approach of an emergency vehicle.

---

6. Thomas F. Adams, *Police Patrol Tactics and Techniques* (Englewood Cliffs, N.J.: Prentice-Hall, 1971), p. 236.
7. Donald O. Schultz, *Police Traffic Enforcement* (Dubuque, Iowa: William C. Brown Co., 1975), p. 48.

A patrol vehicle using a siren should pass traffic on the left even though it is necessary to drive across a center line of the roadway. A motorist will often swerve to the right at the approach of an emergency vehicle. Therefore, an officer should cautiously pass on the right *only when no other course is open*. Sudden use of the siren immediately behind another vehicle should be avoided as it may cause an excited motorist to stop so abruptly that his or her car will be struck by the patrol vehicle.

The purpose of the red light and siren is to minimize traffic delay. Speed above the posted speed is seldom necessary or justified. The red light and siren are not protective armor; the officer can still be involved in an accident, never reaching the scene of the emergency.

The following safety rules during emergency vehicle operation should be observed:

1. Do not follow immediately behind another vehicle and then suddenly blast the siren. The motorist may rapidly apply his or her brakes.
2. Do not pass on the right when using the siren unless no other course is available and then only when reasonably sure that the car being passed will not drive to the right as you go by.
3. Do not sound the siren at its highest pitch continuously; fluctuate its sound throughout the tone scale.
4. Do not lead or escort an emergency vehicle.
5. Drive with due regard for the safety of all persons using the highway.
6. When driving at high speed with red light and siren, keep near the center of the roadway so oncoming vehicles can see the red light approaching.
7. Increase speed between intersections and approach intersections with extreme caution. Use of second gear is sometimes of benefit at speeds less than 75 mph.
8. Make a habit of using your safety belt at all times.
9. The law gives you the right-of-way when the siren and red light are in operation, but you must give motorists and pedestrians the opportunity to yield to you.

10. Check your beat first each day for changes and hazards. An intimate knowledge of your beat and your (and the car's) limitations are to your advantage.

11. Do not be ashamed to ask for assistance. We are a *team*. Radio gives us mobility. Use it! Advise your route, speed, and describe your suspect and vehicle.

12. Maintain a safe distance when observing for violations of following an emergency vehicle too closely to the rear. A motorist may yield to the emergency vehicle and, after it has passed, pull directly in front of you.

13. If involved in an accident, notify your sergeant and do not talk unnecessarily.

While the law provides some legal protection, the officer will find that real protection lies with his or her personal judgment and defensive driving habits, rather than a reliance on emergency equipment. A large portion of the driving public can hardly cope with normal driving problems; we therefore take on added responsibility and must use extreme care when operating an emergency vehicle.

The officer will perform his or her duty best if his or her driving is based upon the following principles, proven by experience to be sound:

1. Get there as quickly as possible with safety.
2. Keep your speed at a level which will enable you to avoid hazards which you anticipate by being alert and exercising due care.
3. Keep your vehicle under control at all times.
4. *Allow motorists and pedestrians an opportunity to yield the right-of-way.*[8]

Auto Theft Investigation

The economic loss due to auto theft would stagger the imagination of most individuals. Each year the unrecovered property loss of automobiles reaches figures in excess of 60 million dollars. Over half the vehicles stolen are from private residences, apartments, or streets in residential areas. Three-fourths of these thefts occurred at night after 6:00 P.M. Only about 9 percent of the vehicles stolen are for commercial purposes, and nearly 64 percent of the

---

8. *Enforcement Tactics Manual*, pp. 71-75.

autos stolen each year are by persons under eighteen years of age. Statistics have further shown that about 55 percent of America's auto thefts are the direct result of the key being left in the car or the ignition left unlocked. Of all the vehicles stolen about 87 percent are actually recovered. What does this mean to the traffic officer or patrol officer? It means that there is a great likelihood that an officer will stop someone in a stolen vehicle at least once or perhaps twice in his or her career. It is therefore imperative that the officer understand some of the basic characteristics of the auto theft.

Strict traffic law enforcement is the most effective weapon with which the officer may combat the auto theft problem. Often a violation committed by a criminal fleeing the scene of a crime, or by a thief driving a stolen vehicle, will provide an officer with the opportunity to stop the car and apprehend a wanted person who might otherwise escape detection. For this reason, the officer must remain constantly alert for the minor violations and irregularities which will lead to an interrogation of the driver and a thorough examination of the vehicle. The more vehicles stopped and drivers checked, the greater the chance of apprehending an auto thief or other wanted person. The following signs should be watched for and investigated:

Damage Indicating Forced Entry. Broken wind wings, damaged door handles or locks, damaged trunk compartment door, ripped convertible top, and a wired ignition system are all irregularities which may indicate that the vehicle has been stolen. Vehicles displaying these signs should be stopped and the driver questioned closely regarding the reason for such damage.

Suspicious Behavior of Driver. Many drivers will display symptoms of nervousness or odd behavior upon being stopped by a policeman. This is a normal reaction and should not be interpreted by the officer as meaning that every driver acting in an unusual manner is a criminal. However, there are certain actions which should arouse suspicion.

1. Extreme nervousness or bluster.
2. The driver who, upon being stopped, jumps from his or her car and hurries back to the patrol car. There may be a reason for not wanting the officer to see the inside of the car.
3. Drivers who attempt to divert the officer's attention or give evasive or hesitant answers to routine questions.
4. Drivers or occupants who do not fit the car, such as poorly

dressed persons in a new car, or exceptionally young persons in a late model family-type car.

5. Drivers or occupants who constantly watch the officer.
6. The driver who attempts to evade arrest for a minor violation.
7. An extremely reckless driver.
8. Drivers who turn into the first available side street when being followed by a patrol car or when approaching one from the rear.
9. The driver who does not seem familiar with the operation of his or her car.
10. Driving without lights.
11. Cars stuck off the roadway due to unfamiliarity or lack of skill on the part of driver.

Registration and License Plate Irregularities. The registration card on all vehicles with which the officer comes in contact should be checked for alterations, erasures, or other discrepancies. All such irregularities demand an immediate investigation, including a careful examination of the engine and/or vehicle identification number.

The license plates will often reveal valuable information to the observant officer.

1. New plates on an old car or old plates on a new car.
2. Plates loosely attached with wire or cord.
3. One plate attached over another.
4. Plates dirty, obscured, or bent.
5. Plates altered through the use of black tape or paint.
6. Plates on front and rear that do not match.
7. One plate.
8. No plate.

Investigation of Suspected Stolen or Wanted Vehicles. A careful search should be conducted of all abandoned vehicles and those suspected of being stolen or wanted. The extent of the investigation and search will depend upon the circumstances present at the time and place that the vehicle is located. However, the following steps can normally be followed and will often reveal important information to the officer:

1. When a car is suspected of being stolen and is licensed in a state where the license stays with the vehicle, close attention should be paid to the holes in the plate normally used to affix the license to the vehicle. If the plate shows that it has been attached to other holes, in addition to the ones being used at the time of observation, there is a strong possibility that this license has been attached to some other vehicle. This is easily detected in a distinct marking around the holes previously used. In states that require the license to remain with the original owner, who has several cars throughout the year, the plates will have been attached and detached several times.

2. It is always good practice to check all cars on which the license plate appears to have been disturbed. The officer should be especially aware of late model vehicles when the license plates are attached with wire or material other than the normal bolt. In many cases, the suspect has stolen a license plate and attached it to the stolen vehicle by means of a piece of wire. Special note should be made as to the location of the plate on the suspected vehicle. In haste, the suspect may attach the plate to the grill of the vehicle or other locations not commonly used by the average car owner. This is an outstanding clue on new cars, since it would be unlikely for the dealer to attach the plate to any place on the vehicle other than the one designated. The vehicle may also be suspect if the old plates or dirty plates have new and clean bolts used to attach them to the vehicle.

3. A broken wind vent may be a clue in determining whether a vehicle is stolen, as the thief frequently uses this means of entry into a locked vehicle. The glass is generally broken near the lock release button and the hole is often plugged with cloth or cardboard. After theft, in cases where the keys are not available to the suspect, the only means of entry to other areas of the vehicle is by force. Close note of the trunk lock should be made. This is especially true with new cars. The punched-out trunk lock may lead to the arrest of a car thief and the recovery of the car.

4. Not often is the dealer's nameplate a factor, but it certainly

should not be overlooked. Sometimes a suspect will remove the dealer's nameplate from the trunk lid to disguise the stolen automobile. This is done by removing the nameplate completely, leaving nothing but noticeable holes in the lid. The suspect may also attempt to hide the nameplate by the use of a can of spray paint, blending it in with the body of the vehicle. In regard to new vehicles, this nameplate should, in most cases, indicate the town or city of the dealership. Generally, this location will not conflict with the state license plate.

5. The service record, usually located on the door casing, will sometimes lead the officer to the origin of the theft if he or she does not know where the car was stolen. Most of these tabs bear the name of the city where the gas station is located and the service was performed. This can be used to question the suspect as to when and where he or she had the car last lubricated, how many miles ago, or approximate date, if he or she insists the vehicle is his or hers.

6. Check the radiator for heat and the surface of the ground near the tires for evidence that would indicate how long the vehicle had been parked at its present location.

7. Are the doors locked or the ignition key in the switch? Check under the hood and dashboard for a jumper wire or similar device.

8. Check I.D. numbers for evidence of alteration or tampering.

9. Check the outside of the vehicle for damage or blood stains which would indicate a hit-and-run accident.

10. Check the inside of the vehicle for blood stains, burglary tools, firearms, narcotics, or other evidence which may indicate its use in the commission of a crime.

11. Check for possible mechanical failure. This may be the reason the vehicle is unattended.

12. Search the vehicle thoroughly for any evidence that would indicate the identity of the owner, occupants, or driver. Letters, envelopes, match book covers, road maps, lubrication tabs, newspapers, and other seemingly unimportant material can often provide the evidence necessary to solve a vehicle theft or other crime. This search should include the trunk

compartment with special care centered under and behind the spare tire. The absence of a spare tire or other accessories, especially on a late model vehicle, will often indicate that the vehicle has been stolen, as a thief without funds will usually sell them at the first opportunity. The glove compartment, under the dash board, and the underside of the seats and floor mats should all be closely examined.

13. Request registration and stolen information providing both license and vehicle identification numbers.

If the vehicle is parked well off the roadway, or in a driveway or in front of a residence, and the doors have been locked and the ignition key removed, it often indicates the driver is nearby and will return soon. If the vehicle under observation is believed to be stolen, or the driver wanted, a stake-out should be set up to await the suspect's return, first taking the precaution of removing the spark plug wires.

If the investigation reveals that the vehicle is stolen or wanted, and the location or the length of time it appears to have been abandoned indicates that the driver is not going to return, or if a stake-out has proven fruitless, the vehicle should be impounded with care taken to preserve any evidence or fingerprints that may be present.

If there is no stolen report and no evidence that the abandoned vehicle has been involved in another crime, it should be observed at regular intervals by the beat officer and removed to a place of storage after a reasonable length of time has elapsed. It is a good policy to check for a stolen vehicle a second time if the vehicle remains in the same position for longer than twenty-four hours, as the stolen report may not have been filed at the time of the first request.[9]

Summary

Tactics used in traffic enforcement may vary from agency to agency and state to state, however, they all have the objective of making the streets and highways safe. This is one of six objectives that should be implemented. The other objectives are to eliminate actual or suspected opportunity for wrongdoing, to regulate traffic, to provide the many necessary services that will aid the motoring public, to prevent accidents, and to determine the problems an officer faces on the beat.

---

9. *Auto Theft Manual* (Sacramento: California Highway Patrol, 1969), pp. 32-40.

All traffic enforcement units carry out various functions such as the enforcement of traffic laws and regulations, control of the flow of automobiles and pedestrian traffic through the means of direction, and the investigation of traffic accidents. Other traffic duties may be performed that are essential for advancing the objectives of an officer's particular department traffic control program.

In order to have efficient patrol, officers should be required to keep up with constant changes in the law. Enforcement methods require that officers maintain reasonable on-view patrol. Observation techniques should be employed so that officers can readily view accident-causing violations and thereby take the necessary enforcement action.

The quality and quantity of enforcement must be effected by good planning. Communication with the public and education provided for them is important. Selective enforcement is predicated upon the fact that the people must know the law; that the law is everywhere. A citizen should know that if he or she breaks the law, he or she will be caught. If apprehended, he or she will be arrested and upon adjudication will be prosecuted.

Officers alert for those accident-causing, moving violations should take some type of immediate enforcement action. Enforcement may be in the form of a written notice to appear (citation or summons). It may be a written or verbal warning and even perhaps a physical arrest. The officer must use sound judgment when taking any type of enforcement action. He or she should be courteous, but firm, and explain the violation.

The quantity of enforcement also varies. A good guideline to follow is one enforcement contact or action for every hour worked. It is a known and proven factor that when citations are up in a critical area, traffic accidents are down. How can we determine just how many citations to issue? This is usually a departmental problem; however, through the use of good selective enforcement over certain designated periods of time, guidelines can be established. Selective enforcement requires that an officer first determine three things: the time of day that accidents are the highest, the locations where accidents are occurring most frequently, and the types of violations that are most prevalent. When all three of these have been determined the officer can then apply direct, selective enforcement to those problem areas.

We looked at the various problems dealing with auto theft and explored some things to watch out for and techniques of investigation to be applied. In order to cover fully the spectrum of enforcement, the next chapter deals with enforcement psychology. The student should also consult the portion dealing with Establishing Traffic Accident Violations in chapter 10. These are good tools to aid the enforcement officer. When properly applied they can work to effectively reduce the accident problems in a community.

DISCUSSION QUESTIONS

1. Identify the various methods presently used by law enforcement to reduce traffic accidents in your city.

2. List some possible guidelines that a police agency might use in developing a good traffic enforcement program.

3. Discuss the positive and negative values of selective enforcement.

4. Identify the various techniques used by auto thieves in stealing motor vehicles.

5. What are the methods used to locate and identify a possible auto theft?

6: The Traffic Stop: Some Psychological Considerations

The purpose of this chapter is to help police officers recognize some of the factors, both in themselves and in the people they stop, that contribute to whether the traffic stop is a positive, neutral, or negative learning experience for the driver.

A positive learning experience is the ideal. The experience is positive when the driver leaves the stop (1) knowing exactly why he or she was issued a citation; (2) aware of the rationale behind the vehicle code section that he or she violated; and (3) firmly resolved to avoid the same mistake again.

A neutral learning experience is probably more typical. Here the individual intellectually recognizes that he or she broke the law, half understands why he or she should not have done so, and half thinks the officer really should have had better things to do. This will motivate the driver for a while, but, usually within a year or two later, he or she will be cited for the same violation—or be injured or killed committing it.

A negative learning experience is one in which the driver is no more prone to avoid the particular violation than he or she was before the traffic stop. This can occur in either of two ways. The officer may mishandle what could have been a positive learning experience and turn it into a negative one; or, the driver's attitude is so defensive and negative that it erases even the best efforts of the officer to make the stop a learning experience.

The results of a negative learning experience are (a) the person "learns" that officers are "stupid" or "fascist"; (b) the person learns that when he or she commits the violation again, he or she should make sure there are not any officers around; and (c) the driver will contest the citation in court. Most negative learning experiences from traffic stops can be prevented if the officer is aware of some psychologial factors that significantly influence the effectiveness of a traffic stop.

Importance of the Traffic Stop

The traffic stop is one of the most important contacts a citizen has with the police for two reasons. (1) It is quite possible that this traffic stop is the first, and perhaps the only, official contact the violator has with the police department. It is often difficult for the police officer to realize this because it may be the fourth citation he or she has issued for the shift, the twentieth for the week, the eightieth for the month, and the two hundreth for the year. The officer will not remember the traffic stop beyond dinnertime. But, for the citizen, this meeting may leave an impression that may remain with him or her for life and may be passed on to his or her children. Even if an officer is not concerned with whether or not people in general like police officers, a citizen who is pro-police is going to make the officer's work much easier, and this attitude should be encouraged by an officer's behavior.

(2) A traffic stop is an opportunity to teach a driver a valuable lesson that may save him or her or his or her family from serious injury or death. To teach anyone a lesson three factors must be present: (1) there must be a "learning atmosphere" which is positive, or at least neutral; (2) the person teaching must be clear and succinct; (3) the learner must be motivated to learn. The police officer can do a great deal to increase his or her skills in the first two areas, and success in them can significantly influence the third factor.

Definition of the Traffic Stop

A traffic stop takes place when a police officer stops a motorist who has committed a violation and issues the motorist a verbal or written warning or a citation. The traffic stop can be divided into three main stages: Stage 1—Pre-Interactional; Stage II—Interaction; Stage III—Aftermath.

Stage I—Pre-Interactional

This stage covers the time period between the moment when the driver realizes he or she is being stopped and the time when the officer and the motorist come into personal contact. All things being equal, this is the most important stage because it will determine to a large extent what will transpire through stages II and III.

It is well known that our behavior toward others is largely based on our perception of ourselves and our perception of the other person.[1] From this flows our tone of voice, the content of what we say, our posture and overall demeanor. In general, there are six "perceptual stances," three of which pertain to the police officer involved with a traffic stop and three which pertain to the driver.

1. The Professional Police Officer. This officer's basic philosophy of police work is that the vast majority of people are law-abiding citizens. He or she is a realist, however, and recognizes that some citizens do break the law.[2] He or she knows that some people break the law unintentionally; e.g., most drivers are unaware that a "U" turn in a business district is illegal. Some citizens may break a law half-intentionally. The individual may vaguely remember reading somewhere that such a "U" turn is illegal, but, since there is no other vehicle in the immediate area and a friend is waiting for him or her in the rain, the driver decides it is all right to make the turn. A few citizens violate the law intentionally; they know the "U" turn is illegal, but their attitude is "So what, it's a stupid law, and I'm in a hurry."

Hence the professional police officer attempts to discern to a reasonable extent exactly what was in the driver's mind so that he or she can exercise reasonable discretion in effecting a positive learning experience for the driver.[3] The officer's attitude is fair and understanding, but firm. His or her demeanor will directly reflect his or her attitude toward himself or herself and others. His or her voice will be calm. The content of what he or she will say to the citizen will be respectful, and his or her questioning will be an attempt to understand and not to entrap the individual. He or she is not afraid to smile or to listen, at least for a reasonable amount of time. His or her posture, though always alert, is not overbearing or combative.

2. The "Lone Ranger" Syndrome. This officer likes to ride off in a cloud of dust, leaving people with awe and admiration. His or

1. James Coleman, *Personality Dynamics and Effective Behavior* (Palo Alto: Scott, Foresman, 1960), p. 68.

2. Michael E. Cavanagh, "Selection of Police Candidates: Mental Health Considerations," lecture presented to Police Officers Standards and Training Seminar, St. Helena, California, January 25, 1972.

3. Theodore Newcomb et al., *Training Key: Volume 1* (Chicago, Ill.: International Association of Chiefs of Police, 1969), p. 11.

her dynamics are based primarily on a personal need to be liked by all, or at least by the vast majority of people with whom he or she comes in contact. He or she rationalizes the "good guy" behavior by convincing himself or herself (and other officers) that police officers have a bad image and someone has to balance it out. The problem here is that the main object of this officer's work is to be liked—not to enforce the law and arrest or cite violators when necessary.

Drivers will leave the encounter with this officer feeling good—or perhaps "relieved," because they did not receive a ticket. Their relief, however, is much like that of school children whose teacher cancelled classes for the day. They "feel good" about the teacher, but they have not learned anything that may be helpful to them or to others in the future.

This officer's demeanor is immature; his other speech is punctuated with "gosh" and "gollies," and he or she laughs a lot. He or she self-consciously shifts from one foot to the other, often not looking at the person he or she is talking to. Immature citizens will think the officer is a "swell cop"; more intelligent and mature people will not be impressed by him or her.

3. The "Wyatt Earp" Syndrome. This officer views most people as "creeps"—rich people because they think that they own the police department; poor people and minorities because they do not like cops and commit most of the crimes; middle-class people who violate the law because they are "dumb bastards"; and educated people because they are "liberals." This "creep list" includes everyone but those police officers who think the same way he or she does. This officer takes as a personal affront any violation of the law in his or her territory (beat) and enjoys playing games of "cat and mouse" with potential violators. His or her world is peopled with "good guys" and "bad guys," and anyone who commits even the worst minor violation is automatically more bad than good. This perceptual stance will automatically and unconsciously be reflected in his or her demeanor. His or her voice will be a bit loud, strained, and caustic; his or her facial expression will be severe, angry, or sneering. His or her appearance includes feet spread apart and thumbs hanging from his or her gun belt. The content of what he or she will say ranges from accusatory to aggressive to combative—even before the driver has opened his or her mouth.

In summary, if one knows which "perceptual stance" best describes a particular officer, one could predict with 80 percent accuracy what kind of learning experience will stem from any traffic stop the officer makes—without knowing any other circumstances of the traffic stop!

4. *The Law-Abiding Citizen.* This driver may be truly law abiding. He or she may have committed a vehicular violation because he or she was unfamiliar with the area or because he or she was very distracted. He or she views the police as a constructive element in society and has always cooperated with them. His or her reaction to being stopped may be one of acceptance of the fact that he or she did break the law and is willing to pay the consequences. More likely, however, the driver will feel that he or she is being treated unjustly because he or she did not intentionally violate the law. If this is the first time that he or she has ever been stopped, the driver will feel he or she deserves only a warning.

5. *The "Let's Be Reasonable Officer" Driver.* This individual also protests that he or she is a law-abiding citizen, but, by this, he or she means he or she has never committed a felony. He or she may violate the same stop sign every day going to and from work, but he or she pays taxes, supports the Police Athletic League, has a brother who is a cop, and is good to his or her family. Hence, he or she is obviously taken aback to be stopped by a police officer for an alleged violation. He or she may be inclined to present all his or her excuses for committing the violation and appeal to the officer's "reasonableness" as a way of avoiding a citation.

6. *The "You Can't Do This to Me" Driver.* This driver, although he or she would not admit it, perhaps even to himself or herself, feels that he or she is immune from the law dealing with misdemeanors. Usually this attitude stems from one of two situations.

The individual may be upper middle class and is professionally successful: an attorney, surgeon, or vice-president of a corporation. He or she is used to giving orders and not used to being reprimanded. He or she may feel it is nonsense that he or she should be stopped, much less cited, for a traffic violation. His or her perception of the police is that they are a necessary restraining force in society and should concentrate their efforts on criminals and not on established people in society.

On the other hand, the individual may perceive himself or herself as a "hassled" member of society. He or she feels that the "establishment" does not like him or her and is trying to force him or her to conform to its neurotic value system. He or she perceives the police as the "enforcers" of the values of a "sick" society and thinks he or she is being cited, not because he or she drove through a stop light, but because of his or her looks or his or her attitudes.

Stage II—Interaction.

This stage covers the period of time from when the citizen and officer first see each other to the point when neither party sees the other again. This stage includes both nonverbal and verbal factors.

Nonverbal Communication. Often, what causes the most problems in any communication is not what is said but what is done. For example, your spouse may keep you waiting an hour and then blithely state: "I'm sorry, I guess I kept you waiting," and go on to explain why she or he is late. But you are so distracted and angry by what her or his lateness communicates to you that you do not hear the explanation. In other words, nonverbal communication can often "shout" louder than words.

The following are cases of nonverbal behaviors and the possible messages they convey, *whether we intend to send the message or not.* Recognizing these, an officer can take steps to either refrain from these actions or, when they are necessary, he or she can explain them verbally to help clear up the confusion or antagonism generated by his or her nonverbal behavior.

For example, when the officer takes an inordinately long time approaching the vehicle after the stop, this creates the possibility of communicating: "Driver, you obviously are in a big hurry; now I'm going to keep you waiting, so you can cool your heels awhile." The officer may have had trouble getting air time to call in the stop, and this is what has delayed him or her. If he or she states this simply to the driver, it would clarify the communication. To give another example, the driver may have heard that it is polite for him or her to alight from the car when an officer stops him or her. The officer, on the other hand, may see that traffic is passing very close to the driver's vehicle; or it may be at night and, since the officer is alone, he or she would feel safer if the driver stayed in his or her vehicle. Abruptly waving the citizen to get back in his or her car could be construed as a hostile, unnecessarily authoritarian gesture unless it is later explained.

Verbal Communication. There are three factors that can greatly influence the effectiveness of a traffic stop.

(1) Tone of voice. When any two people meet each other for the first time, they must rely greatly on external cues for their knowledge of who the other is and what the other is thinking. Con-

sequently, the tone of the officer's voice will convey what he or she is really feeling toward the driver, whether he or she wants it to do so or not. The days are past when it was effective to be the tough cop. One's voice can be confident (versus self-righteous), firm (versus tough), calm (versus unsure), and clear (versus biting). A simple thing like the tone of an officer's voice can transfer what would have been a simple traffic stop into a "hassle" for both the citizen and the officer.

(2) *Verbal dialogue.* The words and phrases we use often reflect our true attitude toward another, even though we may attempt to veil it. When speaking with the driver at the scene of a traffic stop, emotionally loaded phrases are bound to elicit emotionally loaded responses. The following are some examples of statements an officer can make which will increase the difficulty of his or her job and decrease the possibility of effecting a positive learning experience.

"What you did back there wasn't very bright."

"You obviously have no concern for anyone else."

"You almost killed somebody back there."

"For a college kid, you sure don't show it by your driving."

"Maybe you're too old to be driving a car."

"If you weren't so busy yelling at your kids, you would have seen that stop sign."

The most professional framework for the officer to use in his or her speech is the words of the particular vehicle code section that is violated. There are no vehicle code sections regarding "intelligence," "being too old," "not paying attention," or "almost killing someone." If a person failed to stop at a stop sign, the explanation is rather simple: the officer is citing the person for failing to stop completely at the limit line of a posted sign. The more the officer strays from the elements of the violation, the more he or she is opening up the situation for personal invective and misperceptions of what is being said.

(3) *The presence of other people at the stop.* The people in the vehicle with the driver who is stopped probably will have a definite effect on his or her behavior toward the officer.[4] If a man is with a girl friend, or a woman with other women friends, he or

4. Theodore M. Newcomb et al., *Social Psychology* (San Francisco: Holt, Rinehart, and Winston, 1965), p. 211.

she will undoubtedly feel more embarrassed than if he or she were alone. If he or she is not as mature as he or she could be (just as the police officer is not always as mature as he or she should be), he or she may put on an air of bravado in an attempt to salvage or win back his or her feelings of lost esteem.

The most common way of trying to regain prestige is to "put down" the police officer. The driver may say something more or less benign such as: "Society would be a lot safer if you cops were out stopping robberies and rapes instead of hassling people over that lousy stop sign." Or, he or she may say something more poignant such as: "Does this make you feel like a big person—is this how you get your jollies?" There is not a police officer in the world who does not have a repertoire of smart-aleck responses to these typical barbs that some drivers will throw at him or her.

The question is, does the officer whip into a Wyatt Earp role and shoot this varmit out of the saddle, or does he or she perform as a professional police officer by understanding that the driver's remarks reflect his or her own fear and embarrassment and are not personal? If so, the officer can choose to ignore the remarks and attempt to explain the violation and why it was dangerous, all the time not forcing the driver to lose more face in front of his or her friends. Sometimes showing some "class" is contagious. At other times it is not, but the officer can feel good because he or she kept the interaction on a professional level and did not allow the driver to pull him or her away from it.

Sometimes a driver is with his or her family or just with his or her children. This is a delicate situation for two reasons: (1) The officer does not want the driver to look bad in front of his or her family and (2) children are easily impressed and will remember this interaction for a long time. If the driver in this situation is being difficult, sometimes it helps to ask him or her to step back to the officer's vehicle so that the officer can talk with him or her. If he or she refuses, sometimes it helps to have the officer smile at the children as much as to say "Well, cops are not really mean, even though your mother is upset at me right now." It is interesting to see children smile back as much as to say: "Yeh, I know, Mom gets that way sometimes—I'm sorry." If the officer lets the driver get to him or her, then he or she enjoys embarrassing the driver in front of his or her family as a way of getting back at him or her. The only thing this attitude accomplishes is that the officer feels "good" for about five minutes and then tries not to think about it again because it is not very pleasant.

In another frequent situation, the driver is being cooperative, but the passenger (usually a spouse or parent of the driver) begins asking questions or making statements of a belligerant nature. The passenger makes statements such as: "I know he stopped at that sign. Why I was just thinking to myself 'Harold certainly made a beautiful stop at that sign' when all of a sudden I saw your red lights." The Wyatt Earp in the officer is just about to reply "Why don't you keep your mouth shut?" But if he or she can learn to anticipate these situations and realize, beforehand, that this is the friend's or relative's way of trying to make the driver feel better about himself or herself (as much as to say "I love you, even if you go through stop signs"), the officer may take the statement as a challenge, not to his or her esteem or professional competence, but to his maturity as a professional police officer.

If the officer has someone riding with him or her, this will also undoubtedly influence his or her behavior at the traffic stop. If the partners know each other well, there is usually no problem. But if one is trying to impress the other with how tough he or she is, or how sharp or humorous he or she is, or how "he or she doesn't give a damn who anybody is," then the scene is set before the driver even gets out of his or her vehicle.

While a police officer cannot control how many people are going to be witnessing the interaction between himself or herself and the driver, the professional police officer can be aware of these factors and operate in a way so that justice will be done.

Two other problems in verbal communication can affect interaction: irrelevant communication and levels of communication.

Irrelevant communication. Another problem that frequently arises when a traffic stop is made is that the driver, often realizing that he or she is wrong, but wishing to "save face," will bring in other issues that have nothing to do with the reasons for the stop. This will result in statements such as:

> "My brother-in-law's house was burgled last month, and you police still haven't found the crooks."
>
> "By the way, that stop light at Third and Main is stupid and is going to cause somebody to get killed."
>
> "People are always speeding by our house and almost hitting our kids."
>
> "I saw one of your officers sleeping in the patrol car last month."

It is amazing, despite the obviousness of the decoy issue, how many officers "bite" and enter into long, emotionally charged polemics, discussing everything from the competence of the burglary detail to the reasons why police officers carry guns to national issues ("How come with politicians cheating the government out of millions, I have to pay for current tabs on my license plate?"). Decoy issues can easily be fended off by the officer with such remarks as: "I can see you are concerned with that situation. If you'd like to follow it up, this is the person or the agency to contact." There is an important difference between being abrupt on the one hand and finding oneself entangled in a fifteen-minute debate on the other.

Levels of communication. The psychiatrist Eric Berne postulated that all of us have three parts which are necessary for us to function effectively: the adult, the parent, and the child.[5] The adult part of us has three characteristics: it collects data in an objective manner; it estimates the probability that a certain behavior will work or not; and it keeps emotional expression appropriate. The parent part of us is judgmental. It tells us when we are "good" or "bad," and it helps us judge others similarly. If it is healthy, it will keep us out of trouble and make life easier. If it is overbearing, we shall be overly harsh on ourselves and overly judgmental of others, behaving in an authoritarian and condescending way. The child, if it is healthy, helps us behave in spontaneous, pleasurable, humorous, and creative ways. If it is unhealthy, it leads us to behave in "bratty" ways—stubborn, argumentative, selfish, and aggressive.

In general, the most appropriate and productive transaction between a police officer and the driver at a traffic stop is an Adult-to-Adult transaction which is diagrammed in figure 6-1.

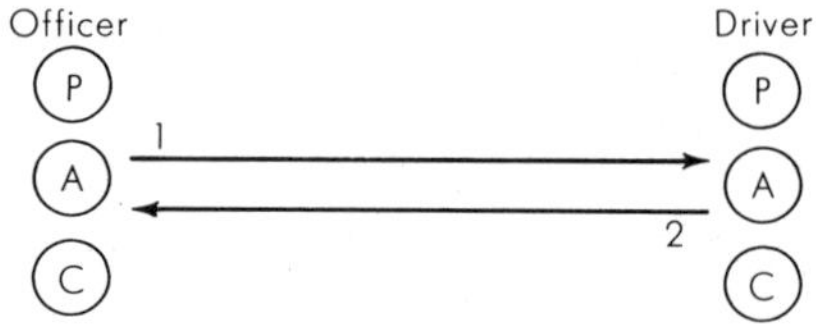

FIGURE 6-1

One example of an Adult-Adult transaction would be:

Officer: "Good afternoon. I observed you going through the stop sign at the last intersection and wanted to call it to your attention."

Driver: Either "I really didn't see the sign. I was so distracted by those boys fighting on the corner that I didn't see it." or "Well, I did see the sign, but figured I slowed down pretty well and

5. Eric Berne, *What Do You Say After You Say Hello?* (New York: Bantam Books, 1973), pp. 11-20.

thought it was safe enough to proceed."

The officer, his or her Adult still "plugged in" with a little healthy Parent added, responds with one of the following statements:

"Well, I saw those kids too, and they were a distraction. Since you don't live around here and aren't familiar with this sign, and since you know you made a mistake, let me remind you that your main responsibility is to keep your eyes on the road because seeing a stop sign and obeying it could save your own life or that of someone else."

"Well, I can understand your thinking on the matter. However, the law does state that a vehicle must come to a full stop, and it's important that citizens can make that assumption of each other. It's for that reason that I am issuing a citation with the hope that it will cause you to remember to stop in the future."

The driver may respond in an Adult way: "Well, I don't like getting a ticket, but you're right. I knew what I was doing and took the chance, so I should take the consequences."

Less ideal transactions, however, can occur. For example, when a police officer "comes on" Parent, he or she can expect one of two types of response. One transaction occurs if he or she engages the Child of the driver, as diagrammed in figure 6-2 and described below.

Officer: "I don't know what's wrong with people like you. You

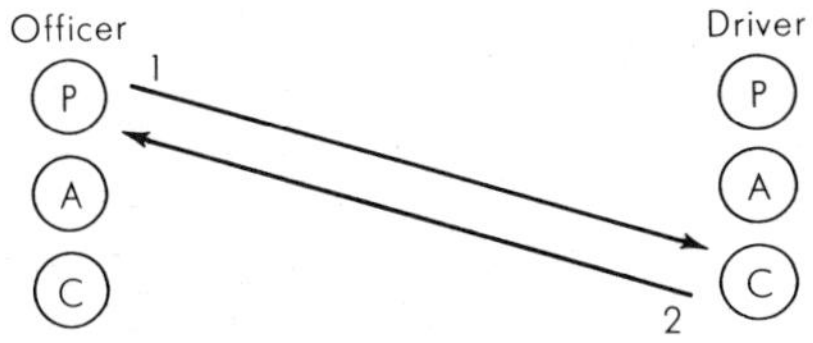

FIGURE 6-2

obviously saw the stop sign and shot through it to beat that yellow Dodge through the intersection. Obviously you feel that where you are going is more important than other people's lives or property."

Driver: (coming on Child) "Big deal—so put me on the ten most wanted list because I missed a stop sign. I guess now you'll get a medal or something."

At this point, the officer will continue his Parental approach, only creating more of a Child-like reaction in the driver, or he or she will switch to his or her Child and enjoy a good verbal skirmish just as he or she did when ten years old.

When the officer in the situation comes on Parent, he or she also may activate the Parent of the driver as shown in figure 6-3.

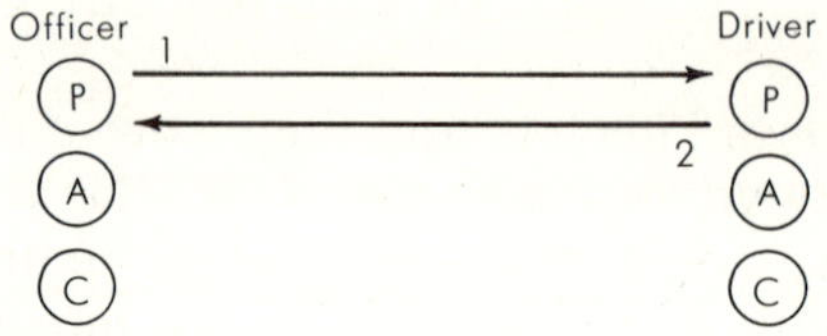

FIGURE 6-3

Officer: "You didn't stop at that sign back there, and being a father with all those kids in the car, you'd think you'd be more aware of what can happen when people don't stop at signs."

Driver: (coming on Parent) "Well, officer, I happen to be an attorney and am not used to being addressed as if I were a delinquent child. I happen to know quite a bit more about the law than you do, and it's too bad you had to bring up my children as an attempt to gain some leverage in this discussion."

At this point, the officer may respond by coming on even more Parental or by plugging in his or her Child to see if that will work better.

The officer may begin a transaction by using his or her Child. In so doing, he or she may hook the Child or the Parent of the driver. This will result in one of the two situations diagrammed in figure 6-4 and described below.

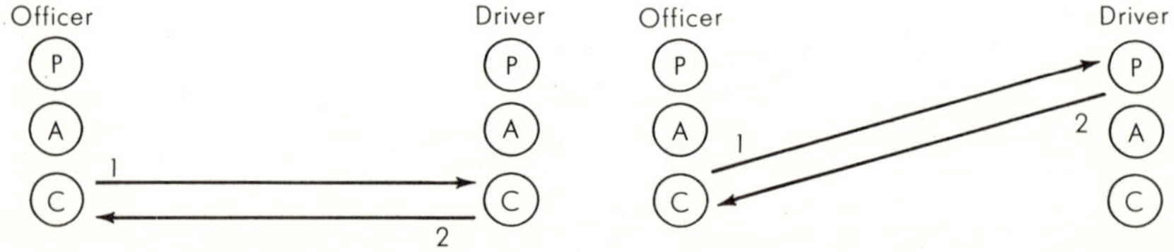

FIGURE 6-4

Officer: "Well, Speed, you just passed eight cars in the last eighth of a mile at seventy miles per hour. You won the race and here's your prize—a nice fat ticket."

Driver: (responding with his Child) "Nice going, old man. It's a wonder you didn't have a stroke trying to catch me. Don't they enforce the retirement age anymore?" Or,

Driver: (responding with his Parent) "It just so happens, officer, that I am a nurse responding to an emergency in the intensive care unit, and if you weren't playing your silly little game of hide and seek, I'd be there by now saving a life!"

In neither case is there much hope for salvaging this situation and turning it into a positive learning experience.

In many situations, the officer starts out well, using his or her Adult. But if the driver responds with his or her Parent or Child, then the officer slips into his or her own Child or Parent. One situation is diagrammed in figure 6-5 and described as follows.

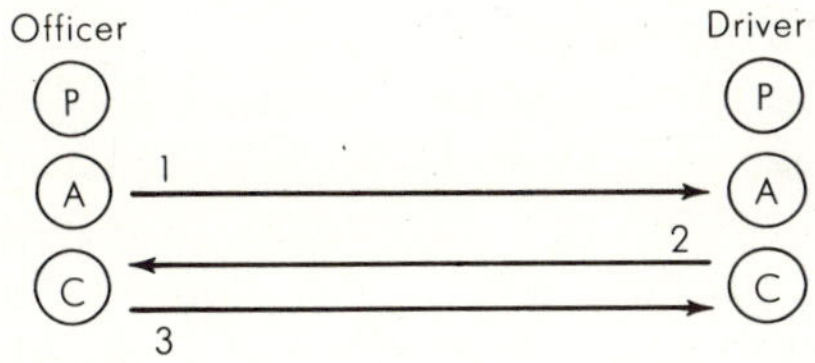

FIGURE 6-5

Officer: (using Adult) "Good afternoon. I observed you go through that stop sign back there and would like to speak with you about it."

Driver: (responding with Child) "Like hell you did. I came to a full stop!"

Officer: (slipping to Child) "Well, I'm glad neither I nor my kids were standing in front of your vehicle when you made your full stop—we'd all be in the morgue by now!"

The continual challenge to the officer is to leave his or her Adult "plugged in" regardless of whether or not the driver is reacting to him or her in Parent or Child ways. The officer, because his or her Adult is operating, sticks to objective data that he or she observed; makes his or her judgment on all the data in the situation; and makes his or her Adult emotional responses to the *situation* and not to the driver. Imagine what would happen if a surgeon allowed his liking or dislike of his or her patient to influence his or her surgery. The officer cannot control completely how the driver will respond to him or her, but he or she can always control how he or she responds to the driver.

Preventing Avoidable Conflict

Two kinds of conflict can occur between people: avoidable and unavoidable. The nature of unavoidable conflict is such that, no matter what is said or done, a conflict will ensue because of the clashing elements inherent in the situation.

Avoidable conflict means that the conflict arises not so much from the situation itself as from the way the person handled the situation. Conflict on a traffic stop, as in any situation, may be entirely unavoidable; or it may be entirely avoidable; or it may be a combination of both.

Most avoidable conflict is caused by an individual's lack of awareness regarding the psychological processes that are taking place both within himself or herself and the person with whom he or she is relating. It is very important for the traffic officer to realize some of the basic psychological dynamics that are particularly relevant to the traffic stop.

The anger-fear syndrome to be discussed in this section applies to *all* human beings. Those individuals who understand how it works within themselves and others are at a marked advantage. The symbolization process which will also be discussed is a part of each human being as well. In some it is mostly a constructive force; in others it is primarily destructive; and in most people it vacillates depending on the situation and the mood of the individual at the time. If the traffic officer can study, discuss, and utilize these basic principles, he or she will become both a more effective traffic officer and a more integrated human being.

The Fear-Anger Syndrome. Fear is the anxious anticipation of danger or pain. This pain may be either physical or psychological. For a traffic officer, the day is replete with the threat of both. On any traffic stop he may be punched, kicked, knifed, shot, or hit by a vehicle.

Psychological pain is often a much more subtle phenomenon, but it can create just as much fear, if not more. It is well known that some people risk death and even choose to die rather than admit to something that radically challenges their self-concept. How many teenagers have been killed in vehicular accidents trying to hold on to their self-concept as being "brave" or "not chicken"? Even adults sometimes choose death rather than have to live with a self-concept that is painful. For example, a person may choose to die to avoid viewing himself or herself as a "traitor"; or another may commit suicide rather than face his or her perception of himself or herself as a failure in business or in love. The point is that psychological threats in traffic enforcement may generate much more fear—mostly unconscious fear—than the threat of physical pain. The direct implications of this will be discussed a little later.

Most normal people share about the same self-concept needs. They need to feel intelligent, reasonable, mature, honest, attractive, logical, emotionally stable, socially acceptable, morally good, alert, effective, and knowledgeable.

On a traffic stop, an officer's self-concept is often challenged or directly attacked. The purposes of such attacks are twofold: to make the driver feel better by making the officer feel worse, and to put the focus on the officer instead of the driver. This is done not necessarily out of maliciousness, but from the citizen's need to protect himself or herself psychologically. We all instinctively do the same thing when we feel we are losing some advantage.

The officer's intelligence is hit ("What can you expect from a dumb cop anyway?"). His or her reasonableness is questioned ("Come on, officer, why be so completely unreasonable about this?"). His or her honesty is assailed ("I'll bet if I were some big politician you'd look the other way"). His or her goodness is denied ("Why are you so hostile and aggressive?"). His or her mental health is ridiculed ("It takes a sick cop to enforce the laws of a sick society."). And his or her competence is doubted ("You were too far away to see if I went through the stop sign.").

When one or more self-concept needs are significantly assaulted, the automatic response of the organism is fear. Fear is a very healthy and helpful emotion because, if we allow ourselves to experience it, it helps us gauge our activities along safe lines. It helps us know what to say or not to say; what to do or not to do, and when to exercise caution. Fear is what keeps people who are in dangerous occupations alive.

Most people in our society have been taught, to a greater or lesser extent: (1) always be strong, and (2) fear is synonymous with weakness or cowardliness. With regard to traffic enforcement, the officer may be in a situation in which he or she is frightened but, because of previous learning, he or she cannot allow himself or herself to feel it. Consequently, he or she reflexively and unconsciously transforms the emotion of fear into anger. Anger is much more acceptable in our society and energizes the person to attack the source of the threat. But this strategy will be more successful in cases of physical assault. When psychological attack occurs, it is the greatest source of avoidable conflict.

Even the best officer can unknowingly convert fear to anger. The more fearful he or she is, the more aggressive or defensive he or she may become. This can manifest itself in many ways. He or she "comes on" a little too "uptight." He or she orders people around in a manner more apt to breed defiance and antagonism than to create order. There is a certain condescension in the officer's manner. He or she may become sarcastic, arrogant, or unnecessarily abrupt as a defense against fear.

A less secure officer who becomes very frightened may become verbally or physically abusive. He or she rationalizes these actions by saying: "You have to keep on top of a situation (people) or you'll lose control." In reality, this too often translates into:

"When you're scared, the best defense is a good offense. If you put people down real fast, they can't hurt you as much."

If an officer can learn to understand his or her fear and realize what is *really* going on between the driver and himself or herself, he or she will be able to prevent a great deal of avoidable conflict. A few simple steps may be helpful.

(1) Realize that what makes a person strong is not an absence of fear but how the individual *handles* the fear. A weaker person recycles his or her fear into anger and attacks. A strong, mature officer recognizes that psychological threat causes some degree of fear in him or her, and he or she deals with it without converting it to anger.

This precludes behavior that will escalate whatever conflict is already present. The mature officer understands that attacking never helps when dealing with a psychological threat. If the driver feels he or she has sensed a weak spot in the officer, it will only be confirmed if the officer retaliates.

(2) Attempt to discern the most vulnerable parts of our self-concept. All of us have at least one "Achilles heel." For some people, it is when their intelligence is assaulted; for others it is their masculinity or femininity; for some it is their maturity; and for still others it is their emotional stability.

(3) Be careful not to jump at the bait that will trigger the threat to the one or two areas about which we are overly sensitive. If a person knows beforehand where he or she is vulnerable psychologically, he or she can protect that part of himself or herself and easily fend off blows without needing to strike back. Remember: no one can *make* us feel or appear foolish without our cooperation.

(4) Realize that an antagonistic driver (this does not include *all* drivers) would like nothing better than to provoke the officer into overreacting, causing the officer to say or do something inflammatory or indiscreet. The driver then has "good reason" to attack the officer and change the focus from his or her own misdeed to that of the traffic officer. The driver feels righteously abused and now can complain to the police department, quote the officer in traffic court, and have a topic of conversation for days. It is *completely* up to the officer if he or she wants to allow the driver this kind of advantage.

The Symbolization Process. Symbolization, in this context, means that when we first meet a person, we have a compelling tendency to relate to him or her, not in terms of who he or she is, but whom he or she *represents* to us. The purpose of symbolization is to diminish the anxiety created by the ambiguity inherent in meeting "an unknown." If we create a "preliminary personality" for the individual (even if it is a negative one), we have some frame of reference—a tentative plan—with which to relate to him or her. Symbolization generally begins when we first see the other person and may continue through the first ten to fifteen minutes of relating. In some cases, it may continue much longer.

The process is almost totally subconscious; therefore, we have to bring it up to our awareness if we intend to successfully work on it. Symbolization can facilitate communication if our symbol of the individual is a positive one; but, it can hinder or even demolish it, if the symbolization is negative. For example, if we have had mostly pleasant experiences with older people, we will meet an older person far differently than if we have had negative experiences. Most people, except those who have become aware of their own symbolization tendencies, approach new acquaintances on the principle: "I've dealt with people like you before, and my past experiences have been pleasant (or unpleasant)." This, in turn, effects our demeanor (tone of voice, facial expression, stance, vocabulary) and creates an impetus that will result in a smooth or conflictual transaction. The ingredients that go into the symbolization process consist of one's overall past experiences with his or her father, mother, siblings, same-sexed and opposite-sexed peers, authority figures (teachers, coaches) and with people of different religions, races, and socioeconomic levels.

Though we confidently assert that we approach all people in the same way, in reality we are truly unaware of the very different methods by which we approach people. Sometimes we catch ourselves saying: "He's not a bad guy, once you get to know him." Frequently, this translates into: "Once I stopped my negative symbolization of him, I saw him for who he really is." Also, comments such as "I was pleasantly surprised," or "It turned out better than I had anticipated" are often examples of our erroneous placement of a negative symbol on a person or situation.

The symbolization process is also self-perpetuating. By this is meant that, if imminent contact with an older person "flashes" a negative symbol for us, our brain automatically "plugs" our demeanor into a "defend or attack" position. The other person almost subliminally senses this and responds to the "invitation" by "attacking" in one form or another. This then "proves" to the first person that all of his or her expectations were correct; namely, that older people (for example) are "mixed up" in some way or another.

The symbolization process is extremely important in traffic enforcement. When a traffic officer and a driver meet at a traffic stop, it is almost always a "first meeting." The instant one party sees the other, the symbolization process begins. Depending upon the positive and/or negative background of each one's symbolization, the traffic stop will have no avoidable conflict, or some avoidable conflict, or be replete with avoidable conflict.

The symbolization process is not an "all or nothing" phenomenon. All human beings have at least latent symbolizations that become activated under certain kinds of stress. The following are examples of potentially problematic symbolization that even good and normal people have. If the traffic officer is honest with himself or herself, he or she may see that, when he or she has had conflictual traffic stops, the majority of them were probably with one or two types of the following categories. With introspection, the officer may be able to gain some insight into what caused such conflicts for him or her and why they happened.

Older men. When older men are negative symbols for the traffic officer, this often stems from the fact that a father or father substitute was or is perceived as cold, rejecting, and punitive, or as weak and frightened. In approaching the vehicle to speak with this driver, the officer may already be promising himself or herself: "I know how this guy is going to come on, and as soon as he does, I'm going to set him straight." No matter how hard he or she tries, there is no way that the officer's preconception of the driver is not going to influence his or her demeanor. If the driver responds to the subtle challenge (even though *he* may be unaware that this is what he is doing), and the avoidable conflict ruins the traffic stop, the officer reassures himself or herself that it was not *his* or *her* fault.

Older women. When older women are negative symbols, this often comes from a mother or mother substitute who is or was seen as domineering, possessive, nagging, emasculating, or weak. When the officer meets an older woman on a traffic stop, he or she may subconsciously say to himself or herself: "Oh, no, here we go again (just like at home)." Instantly, his or her brain "clicks" to the "get ready to be bitched at" position and activates demeanor in like manner. Before he or she is through, he or she probably will get what he or she unknowingly requested.

Opposite-sexed peers. These are people of a sex opposite to that of the officer and who are roughly in the same age bracket (give or take ten years). When a male officer has a conscious or unconscious resentment toward women, it is normally because he has been (or is getting) hurt by or rejected in a romantic relationship with a woman. If an officer feels he has been "burned" by a woman, the traffic stop can afford him ample opportunity to get even, although it is not the same woman that has hurt him.

A woman officer who has trouble reconciling her sexuality with her role as a police officer may also experience avoidable conflict in this area. Because of her past experiences she may view men in this category as "chauvinistic" or derisive of women in police work. Before she has asked to see the man's license, she has said to herself: "This guy is going to try to make fun of me or not take me seriously. So, I'm going to have to let him know immediately that I mean business." Although the officer does not know it, she has just sent out an invitation for the man to challenge her.

Adolescents and young adults. The people in this category are roughly between sixteen and twenty-five years of age. There can be three dimensions to this symbolization process.

(1) The officer who had painful experiences in his or her own adolescence—especially if he or she felt unaccepted as "one of the gang"—may, unconsciously, see this as an opportunity to get back at all the kids that made him or her feel terrible. This seems to be especially true for the officers who are not too far removed from their own adolescence, and for whom the memories are still fresh and bothersome.

(2) Some middle-aged officers may, deep down, resent the fact that they are losing or have lost their youth. Consequently, they resent those who still have it. Such an officer is often heard to make statements such as: "If I had the opportunities that kid has, boy what I would do!" He or she rationalizes his or her behavior as an attempt to "teach these kids to appreciate what they have."

(3) Some officers, because of their own learned values, may seem to be very middle class with regard to their attitudes. But, underneath, they are jealous of the freedom that young people have today. On one level, they scorn and ridicule "these college kids" or "hippies" who sex, drug, and drink at their pleasure, while their parents pay the way or while they live on welfare and food stamps. But, beneath this surface attitude, the officer is jealous of them and wishes that he or she had the psychological and financial freedom to live what appears to be a very carefree and pleasure-packed life. Usually the facial expression of the officer gives him or her away before he or she even gets to the driver's window. When avoidable conflict happens, he or she explains it by assuring himself or herself that the confrontation was due to

the fact that kids "just don't respect authority these days."

The officer can evaluate how constructive or destructive his or her own symbolization processes are. The questions he or she might ask himself or herself are:

(1) Of the symbolization areas mentioned above, which one in particular am I inclined to experience trouble with? Which area "comes in second" with regard to a possible problematic area for me? (It would be interesting to have partners answer these questions for each other and see how much agreement there is.)

(2) Is there something I can do in my private life to defuse this area? Is it possible to try to resolve or at least partially resolve an area of conflict with some significant person in my life that is feeding into my symbolization process?

(3) Knowing the one or two areas that affect me most, what concrete steps can I take prior to approaching the driver to diminish the negative vibrations I am involuntarily sending?

(4) If I do not want to be prejudged by others because I am a police officer, how can I prejudge another on the basis of the kind of vehicle he or she drives, or his or her looks or sex?

(5) When conflict does occur on any particular traffic stop, what part did I play in igniting it? What could I have done to terminate it as soon as it started instead of allowing it to continue and escalate?

Symbolization, of course, is a two-way street, i.e., the driver is also inclined to symbolize the officer. To the driver, the police officer may represent a parent who never had time to love but was always there to punish when a mistake was made. Because the officer is an authority figure (and represents parents and teachers), he or she may instantaneously recall feelings of hurt, injustice, and failure that the driver experienced with overly punitive or perfectionistic parents or with teachers that he or she could never please.

The officer may represent a boss whom the driver feels is ruining his or her life but with whom he or she feels powerless because he or she cannot change jobs. Now, here is the officer (boss) telling the driver he or she made a mistake, but he or she does not have to take a reprimand from *him* or *her!*

An officer may represent all that the driver feels is wrong with our society. The uniform may represent a militaristic attitude; the badge the misuse of power and control; and the gun the reliance on force and the willingness to kill in order to attain society's goals. To some people, the police officer who is giving a citation represents all the real and perceived injustices they have experienced at the hands of society.

There are four important things for the traffic officer to remember in the traffic stop situation: (1) These are not "bad" people, anymore than officers are "bad" when they view hippies as "creeps." People who symbolize are normal, often good, people, who have not found a constructive way in which to deal with past hurts or present fears.

(2) The officer should realize that when a driver verbally challenges or attacks, it is not a personal affront. The driver has never seen the officer before, much less knows anything about him or her. If the officer can distinguish between who he or she is personally and what he or she symbolizes to people, he or she will understand that it is not his or her *person* that is under attack, but merely a symbol someone has "pasted" on him or her.

(3) The officer can refuse to "play out" the role the driver has assigned to him or her. He or she can sense the "negative vibes" or the obvious taunts and say to himself or herself: "I don't know what kind of person this driver is trying to make me into, but I don't want the part. No one is going to coerce me out of being me. If my attitude happens to force these people into reconsidering their image of authority, society, or police officers, that will be fine, too."

(4) When the officer agrees to play the part assigned by the driver, he or she is merely participating in a neurotic complicity. The problematic driver is saying: "You represent half of all that is bad with my life," and the officer is saying, "Yeh, that's right, and you represent half of all that is bad in *my* life!" Both the driver and the officer leave reassured that the world is filled with undesirables. The police officer who handles a traffic stop in this manner has shown no more mental health, maturity, or just plain "class" than the driver that he or she just stopped.

Stage III—The Aftermath

This stage covers the period after all interaction has ceased. Both the officer and the driver have returned to their respective vehicles and have driven away. If the officer successfully effected and completed the traffic stop, he or she and the driver will leave with positive reactions to the encounter.

The driver. Very few people are going to feel good after receiving a traffic citation. "Feeling good" is not a necessary part of a positive learning experience for the driver. If a traffic stop has been handled well by the officer, the driver, in general, should feel:

—that he or she did commit a violation;

—that he or she made a mistake, either by not being alert

enough to what he or she was doing and to his or her surroundings, or by taking a chance that he or she would not get caught;

—that the time he or she tried to save was not worth the chance of possible injury or the time, money, and delay which the citation will cost;

—that if something unforeseen had occurred, he or she might have caused property damage or personal injury;

—that the officer did not make this a personal confrontation but was doing his or her job;

—that the officer provided a nonthreatening atmosphere due to his or her approach, his or her explanations, his or her nonjudgmental attitude, and his or her general demeanor.

The officer. It is ironic how often police officers are dumbfounded that many drivers cannot admit that they made a mistake, despite all the evidence that indicates that they did. But, how many police officers, after a traffic stop, can admit: "I blew it—I shouldn't have said that"; or "I should have done the other." Yet, the only way in which police officers can grow in their profession is to critique their work, admit what could have been handled in a better manner, discover why certain things were done, even though the officer knew better, and look forward to the next similar situation with the expectation that it will be handled more professionally.[6]

At almost every traffic stop there is usually something to learn. The officer who handles all his or her stops mechanically and never gives them a second thought will be making the same mistakes year after year. If the officer returns to the patrol car swearing under his or her breath, throws the citation book onto the seat, grabs the microphone and snarls that he or she is going back "in service"—the officer probably could have handled the stop better, whether he or she can admit it or not. If the officer's excuse is "Well, the driver started it when she said. . .", the officer should become aware that anyone can make a good traffic stop on a driver who is polite and cooperative, either because the driver has a good attitude or because he or she is too frightened to do anything else. It takes a professional to create a positive learning experience with a driver who is defensive or abusive.

The officer who makes a good traffic stop feels good about himself or herself—not because he or she just won an argument, but because he or she taught the driver a positive lesson. The officer feels that he or she has a pretty good job and that most people are pretty decent. The officer is not pent up and looking wild-eyed for the next violator so that he or she can avenge himself or herself.

---

6. Sidney M. Jourard, *Personal Adjustment* (New York: Macmillan, 1963), p. 181.

Summary

The purpose of this chapter is to help police officers recognize some of the psychological factors and interactions that determine whether a traffic stop is a positive, neutral, or negative learning experience. The importance of the traffic stop as the only official contact of the majority of citizens with the police was discussed, as well as the value of the traffic stop as a learning experience for the driver.

The traffic stop is divided into three stages: the Pre-Interactional, the Interaction, and the Aftermath. The Pre-Interactional Stage is primarily concerned with the particular perceptual stance that the police officer and the driver bring to the situation. The Interaction stage discusses various factors involved in the traffic stop itself: nonverbal communication, verbal communication, the presence of other people at the traffic stop, irrelevant communication, and levels of communication. The third stage, the Aftermath, discusses the attitudes with which the driver and the police officer should leave the traffic stop if it has been a positive learning experience.

DISCUSSION QUESTIONS

Comment on the following statements:

1. If a driver is issued a citation, he or she will think twice before committing the same violation again.

2. The main purpose of a traffic citation is to punish a driver for violating the law.

3. By the very nature of a traffic stop, there will be a "winner" and a "loser." If the traffic officer evolves as the "winner," he or she has made a successful traffic stop.

4. The best way to handle an abusive driver on a traffic stop is to "put him or her down good" so that he or she recognizes who is in charge of the situation.

5. A good traffic officer never feels frightened on a traffic stop—he or she knows that there is nothing anyone can say or do that can hurt him or her.

6. The average traffic officer treats all people in the same manner, regardless of who they are or what they say.

7: Traffic Direction

The maintenance of a free and orderly traffic environment is of primary concern to any traffic enforcement agency. Among the functions of traffic control, the importance of traffic direction cannot be overemphasized. Ordinarily, traffic control devices and signals are sufficient to handle most traffic situations. Their primary function is to regulate, guide and warn traffic along with the assignment of right of ways. Occasionally, predictable and unpredictable variables occur which are disruptive to the normal traffic flow, for example, major sporting events, electrical failures, traffic signal failures, traffic accidents, and highway blockages. Events such as these may necessitate manual traffic direction and control.

Signals and Gestures

When an officer has to direct traffic it is his or her job to tell people how, when and where they may move in vehicles or on foot. Therefore, the officer must pay attention to cars, pedestrians, and bicyclists and see to it that they are given a chance to proceed. There is a natural tendency to overlook pedestrians and bicyclists. Actually, what the officer does is tell them how to behave. If he or she were not on hand to make decisions and direct movements when traffic is heavy, drivers and pedestrians would foolishly try to move at every chance. This causes repeated traffic jams. Motorists would also, without realizing it, get into dangerous situations which they did not anticipate.

The officer's most important job while directing traffic is to let drivers and pedestrians know what he or she wants them to do. If they do not understand him or her, they will have trouble and so will he or she. The instructions that follow were developed to facilitate the communication process between the traffic director and the highway user. That is, they show the officer how to make his or her meaning clear to drivers and pedestrians.

Drivers are not often where they can hear an officer when he or she wants them to respond. So the officer cannot just talk to them. He or she has to use a type of sign language which is clearly understood by everyone. The officer could, of course, motion drivers to stop, start or turn in many different ways. Suppose that he or she does it in a way that is entirely his or her own and different from all others who direct traffic. Perhaps some of the drivers approaching the officer's corner would catch on, but most of them would be puzzled. Since they have never come across such motions before they would probably fail completely to understand the directions and ignore the officer. Therefore, it is important in directing traffic for *all officers to do it the same way*. Equally important, the gestures should be made where the driver can see them from a reasonable distance.

Directing Traffic by Hand

When directing traffic by hand the officer must let people know that he or she is in command. In order to facilitate this image the officer must stand where he or she can be seen and as though he or she means business. To combat fatigue and be able to stand for sustained periods, he or she should stand straight with weight distributed equally on each foot. When not signalling his or her hands should hang easily at the side (see fig. 7-1). He or she should not face vehicles he or she has authorized to move but stand with his or her side toward them. This stance conveys a message which tends to be less disruptive of normal traffic movement.

The assignment of right of way to cross traffic is accomplished by first stopping through traffic. To stop traffic the officer extends an arm, points a finger and looks straight at the driver he or she wants to stop. The officer holds this position until the driver sees him or her, then with arms extended the officer exhibits the palm of the hand toward the driver until he or she stops (fig. 7–2). When traffic is flowing in both directions, the officer stops one

FIGURE 7—1
Stance

FIGURE 7—2
Stopping traffic

FIGURE 7—3
Stopping Traffic from Both Directions

side, and then the other (fig. 7—3). He or she should not lower either hand until both directions stop.

To start traffic which is stopped, the officer must decide which side of traffic is to be started first. Then he or she extends an arm and points a finger toward the car he or she wants to proceed. After gaining the driver's attention, with palm of hand facing upward, the officer swings his or her hand up and over to the chin (fig. 7–4). This motion involves bending the elbow. If traffic is to be started from both directions, he or she starts one side, drops the arm, then starts the other side. The officer may emphasize the same signal to proceed for timid or slow drivers.

FIGURE 7–4
Starting Traffic

Right turns require the least direction and signalling. The arm the officer signals with will be determined by the car's direction in relation to his or her position at the time it approaches. When a vehicle is approaching from the right, the officer points toward the driver with the right arm, then swings the arm toward the direction of the intended turn (fig. 7–5).

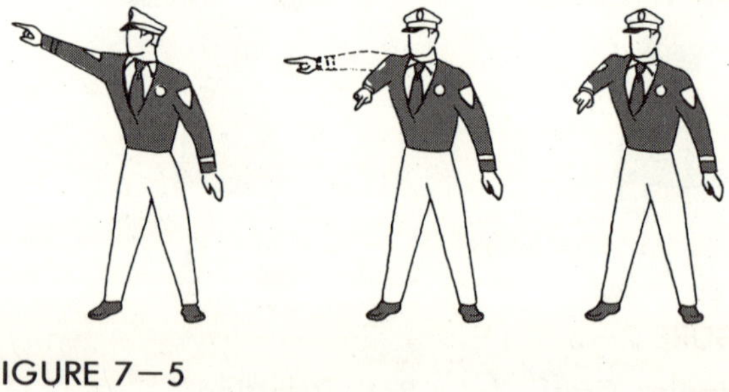

FIGURE 7–5
Right Turns

Left turns require more effort on the officer's part. He or she must be constantly aware of opposing traffic. When appropriate, he or she halts opposing traffic with the right hand and holds it. Then with the left hand, the officer points to the driver desiring to turn left and gives the signal to turn (fig. 7–6).

FIGURE 7—6
Left Turns

Positioning of left-turning vehicles is important, especially when cars are holding up traffic while waiting for an opportunity to turn (no left turn lane available). The officer points to a spot on the highway near him or her where a car can wait, thus clearing the traffic lane for through traffic. This procedure works best when only one car is trying to turn left. When a larger number are present, there may be a need to assign the right of way to the left-turning vehicles.

Two Officers Signalling. The magnitude of a traffic problem may be such that more than one officer is needed to control the situation. When two officers are working together at one location, one of the two must originate all signals and gestures. The reason for this is obvious. When two originate signals, conflicts will occur and occur quickly. The net effect could be total chaos rather than a smooth operation. One officer should always be in charge. It would be wise to assign responsibilities with regard to who will be responsible for each traffic entrant. For example, one officer should direct all north- and westbound traffic including turning movements, and the other all south- and eastbound traffic including turning movements.

Signalling Aids

Signalling aids consist of whistles, flashlights, and voices. The whistle, when used, should make sharp and distinct sounds. The following signals with the whistle are universal: go—two short blasts; stop—one long blast; and attention getter (daydreamer)—several short blasts.

The voice is seldom used in directing traffic. Arm gestures and the whistle are usually sufficient. Verbal orders are not easy to give or understand. However, some talking may be necessary, and if used, the officer must be polite and brief. Simple and concise statements are a must. Conversation should be avoided unless time permits brief explanations to insistent motorists.

Intersection Control

The officer's responsibility is to regulate crossing traffic. He or she should determine whether east-west or north-south traffic will move, and for how long a period of time without interruption. He or she must control turning movements; coordinate vehicle movement at the intersection with adjacent intersections; detour traffic when necessary; supervise signal obedience and, if necessary, direct traffic to disregard signal indications; protect pedestrians and prevent them from illegally crossing the highway; prevent illegal parking and vehicles from stopping at locations that will interfere with traffic movement; provide for the safe passage of emergency vehicles; assist persons seeking information or assistance when time permits; and handle accidents within the area of his or her control until an accident investigation unit arrives. This list of responsibilities is not all-inclusive; however, it is a representative sample.

Positioning at intersections is important for efficient traffic directing. It was pointed out earlier that the officer should stand where he or she can be seen. Some common positions are center of the intersection, corner of the intersection, and center entrance to the intersection (see fig. 7–7).

The best position within a controlled intersection may be any one of these positions. It all depends on the type of intersection and prevailing traffic condition. Within the uncontrolled intersection the center position is usually the best. As far as irregular intersections, multilane intersections and divided highway intersections are concerned, the officer will have to adjust to the situation. More than one officer may be needed to adequately control the intersection.

Any time a position other than the center position is used, a systematic pattern should be developed. One criterion is, ordinarily, that movement should be unidirectional (moving in the same direction to reach each position).

The flashlight is intended for nighttime use. Several techniques have been developed and adopted for effective use of the flashlight. To stop traffic, the officer slowly swings the flashlight at arm's length across the path of approaching traffic. When headlights illuminate the officer, he or she reverts to the regular hand signals and gestures (fig. 7–8). The main purpose of the flashlight is to attract the attention of drivers. Its directional effectiveness is limited. However, when some lighting is present, signalling to keeping traffic moving can be utilized. It must be recognized that a too rapid flashlight movement will decrease its effectiveness.

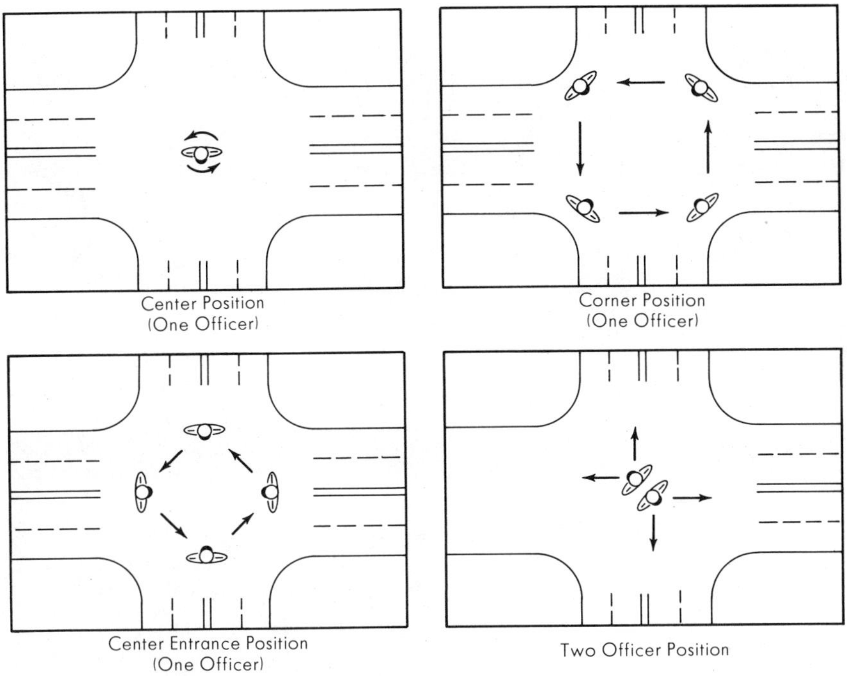

FIGURE 7—7
Intersection Positions

Traffic can be overregulated. Needless regulation will cause motorists to wait to be directed. At signalized intersections it is recommended that traffic be directed in a manner which is familiar to drivers. This means that at a particular intersection, traffic direction should follow the same sequence as the traffic signal. This is recommended when controls are functioning or when they are not functioning. The former is obvious; however, the latter may not be so obvious. For instance, a signal customarily assigns the right of way to the left turner in left-turn lanes first. When the turning vehicles have cleared the intersection, then through traffic proceeding from the same direction is allowed to go. If this is the sequence, then this is the way traffic should be directed.

The rationale for following customary patterns is that drivers are prone to follow patterns with which they are familiar. Directing them in familiar paths helps to facilitate the communication process. Typically, the traffic stream has a high concentration of repeat travelers and the ease in which they follow an officer's direction helps minimize traffic conflicts. Further, no decision is necessary on the officer's part as to how he or she will assign the right of way. If he or she knows the intersection and the signal sequence, he or she simply follows the sequence.

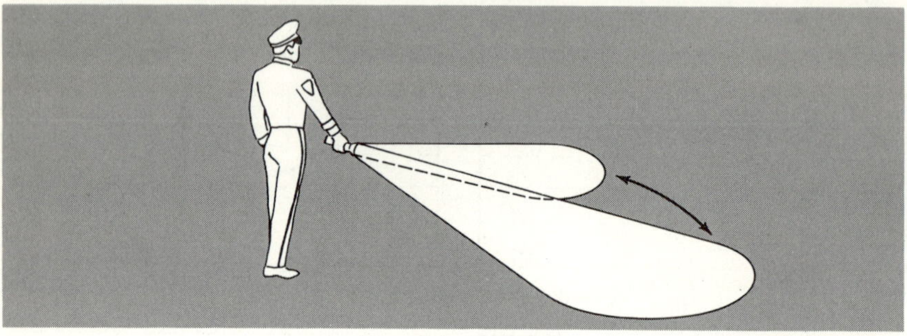

Direct Light Across Path

Use Normal Signals

FIGURE 7—8
Flashlight Use

An officer may, on occasion, direct traffic in conjunction with a functioning signal. The traffic situation is such that the signal is not adequate to control traffic by itself. This may be a high volume traffic situation with frequent demands for left turns with few turning opportunities. In most cases, the signal will do most of the work; however, action on the officer's part will be required from time to time. He or she will be afforded more time to be alert for symptoms of jamming. At the same time he or she must anticipate signal changes.

All exits of the intersection must be watched. If the intersection starts to fill up, remedies to prevent complete blockage must be effected. The officer should check the traffic stream and allow cross traffic to cross when practical. He or she can anticipate congestion and attempt to prevent vehicles from entering the intersection unless they have ample exit space.

The officer must be alert for blockage between intersections. Double parking, unauthorized loading, and minor accidents may be the cause of blockage. He or she determines the reason for mid-block horn blowing and, if necessary, leaves his or her post and clears the obstruction.

The officer can prevent tie-ups by making certain that motorists trying to make left turns do not hold one or more lanes of traffic up. He or she may have to use "no turn" signs or resort to detours.

A list of general rules for smooth operation during intersection control include:

1. Use uniform signals and gestures.
2. Try to break traffic at natural gaps whenever possible.
3. In the absence of normal breaks, try to break the line behind slow moving vehicles such as large trucks.
4. Keep stragglers and "day dreamers" alert and rolling, and in their proper lane.
5. Do not get excited.
6. Do not leave a position just to bawl out a driver.
7. Look cheerful; be cheerful, but be firm.

Flare Patterns

Highway flares (fusees) are used as a temporary measure to control traffic for a relatively short period of time until the situation requiring their use can be corrected, or until other devices have been installed and maintained by the appropriate agency having responsibility for maintenance of the particular highway. There are many events which may occur requiring the need to warn oncoming traffic. The most prevalent situation is the traffic accident.

When an officer arrives at an accident scene, there are several acts which are required. First is the need for an immediate assessment of the situation. The officer should visually check the scene and take notice of hazards and blocked lanes. He or she decides whether a flare pattern is needed, and if so, what pattern to use. Occasionally, passing motorists will have placed flares at the scene. If flares are already placed on the roadway upon the officer's arrival it may be necessary to rearrange the pattern to get the desired channeling. Assessment of situations will aid in the establishment of priorities. The handling of emergencies according to relative importance is essential.

When two officers are present one should immediately establish a flare pattern if the obstruction cannot be readily moved and the other should handle other emergencies at the scene. A single officer should immediately establish a flare pattern unless first aid is urgently needed. In some situations he or she may have to utilize the aid of the public. Usually, they are very willing to help. When the officer is alone at the scene he or she places flares in a criss-cross pattern or stack so that when the first flare has nearly burned out, it will light the next flare (fig. 7–9).

Thought should be given to the proper location of flares to give approaching drivers a chance to slow down before coming upon the obstruction, and to channel traffic into a safe path around the accident. Where the officer places the flares will vary depending upon the accident, the roadway, weather conditions, and what the officer wishes to accomplish. When investigating accidents near curves and crests, he or she should always put flares out on the other side of the curve or crest to warn the oncoming drivers that a hazard is present just around the corner or just over the hill (see figs. 7–10, 7–11, 7–12, and 7–13 for flare positions).

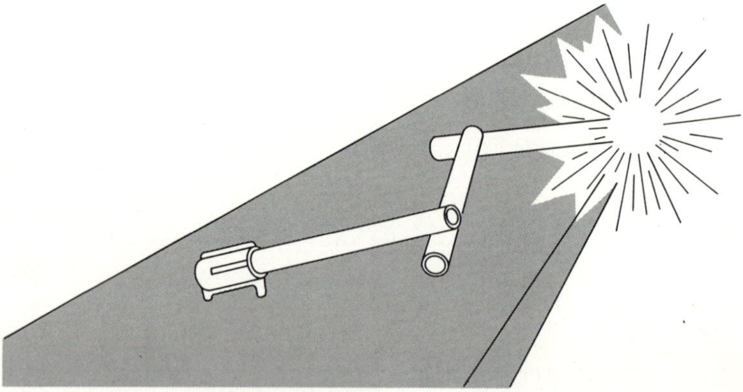

FIGURE 7–9
Stacking of Flares

It should be noted that there is just as much danger in having too many flares placed at a scene as too few. More flares than necessary will cause blending to occur. The red glow of the lighted

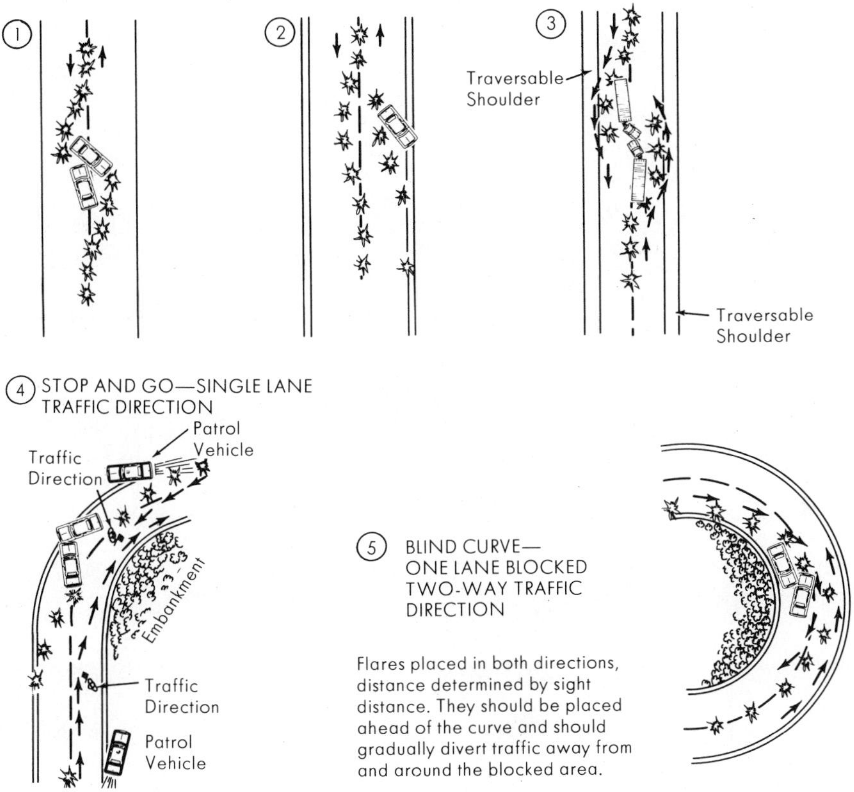

FIGURE 7—10
Two-Way Roadways Flare Positions

flares will blend in with all the flashing red taillights of stopping vehicles and will lead to confusion. Flares set too close together from a distance will blend into a single light and lose their value. Therefore, flares should be placed in a straight line at least 20 to 25 feet apart nearest the obstruction with a gradual increase of 50 to 100 feet apart at the farthest distance needed.

Flare patterns should be set to direct traffic to one side only and should lead traffic away from the wreckage and injured. Decision-making situations should be avoided. Traffic should be led in simple paths and straight lines. When it is desired to have traffic change lanes, the officer sets a gradual alignment which will accommodate the speed of passing traffic.

In determining how far away to start a flare pattern, the legal speed and the actual speed of vehicles using the roadway must be

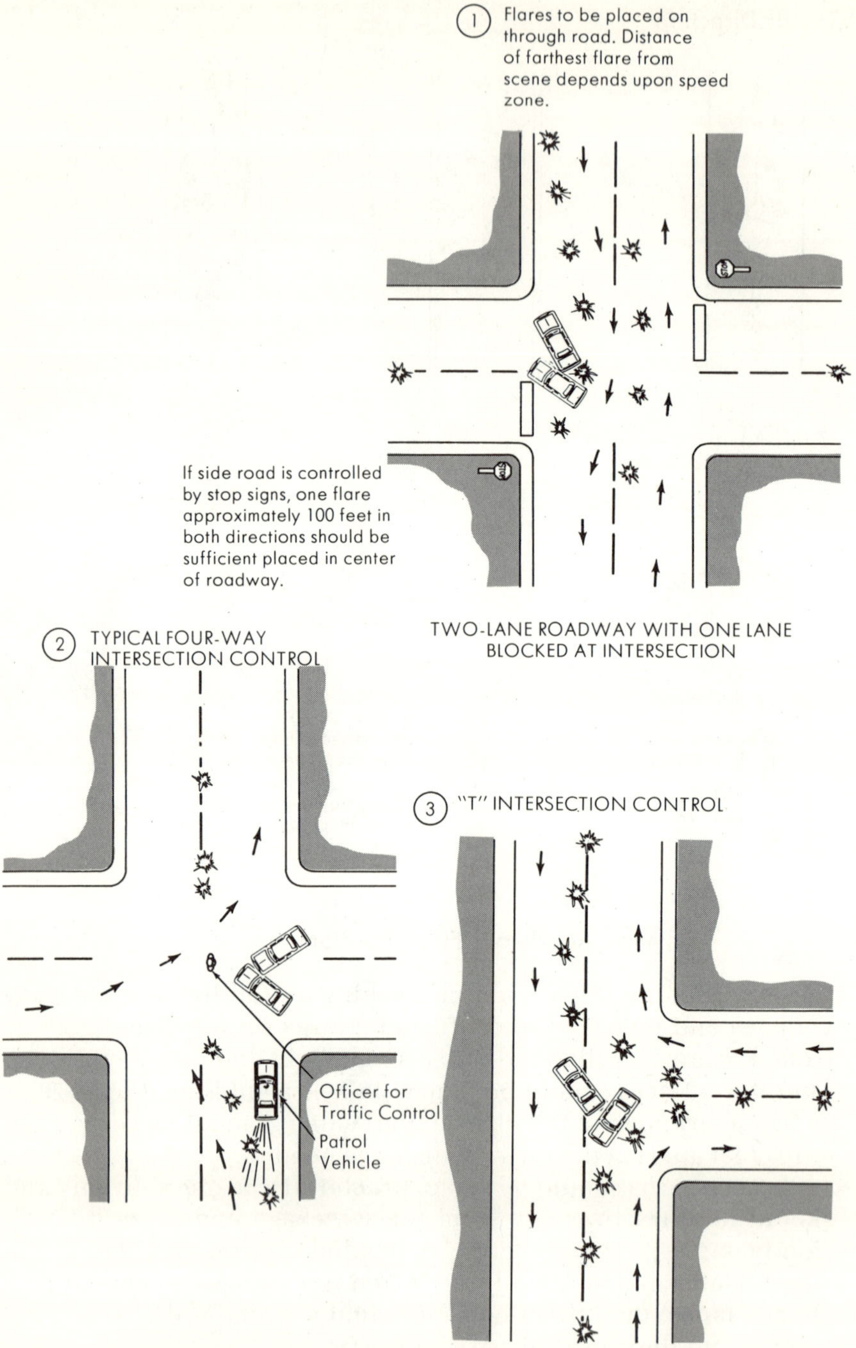

FIGURE 7—11
Intersection Flare Positions

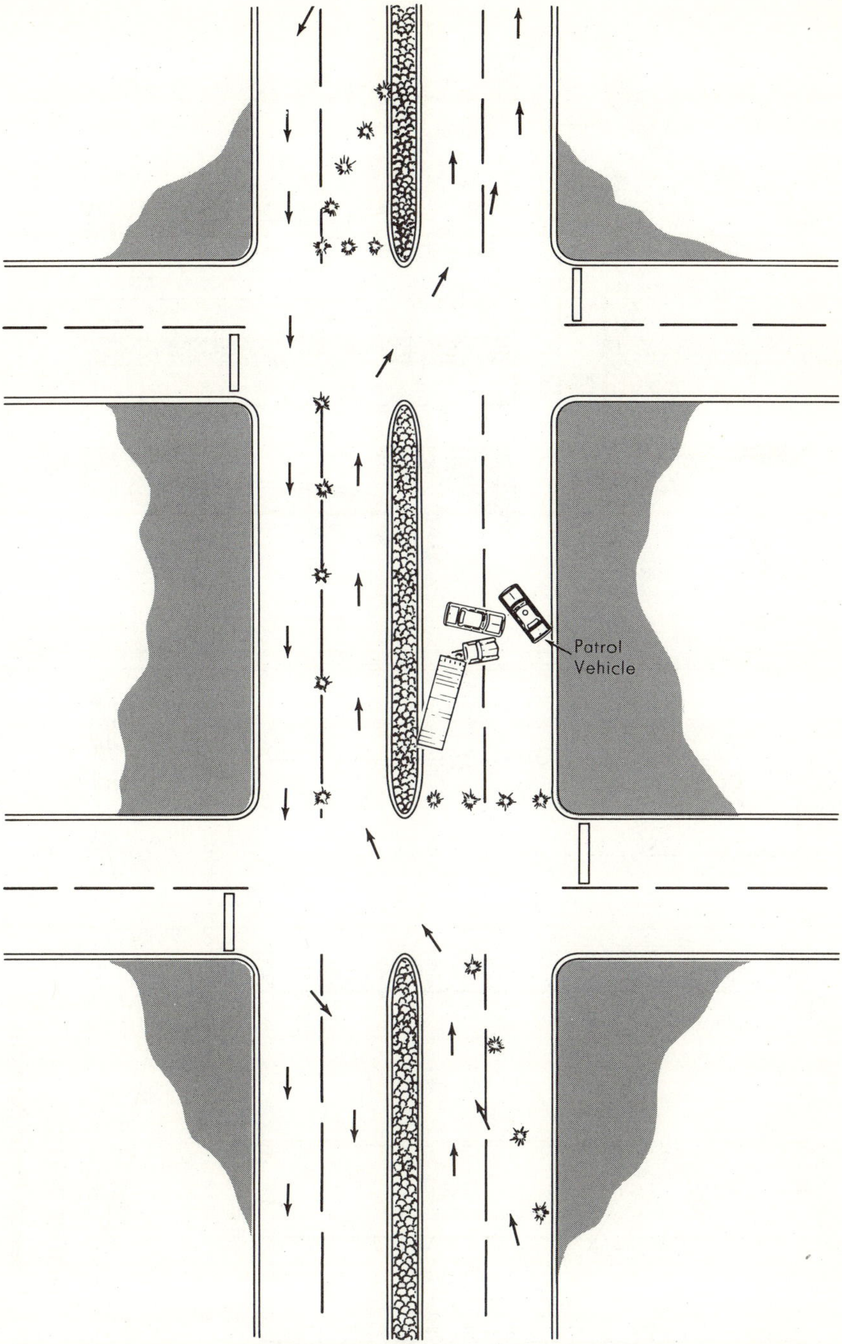

FIGURE 7—12
Expressway Flare Position

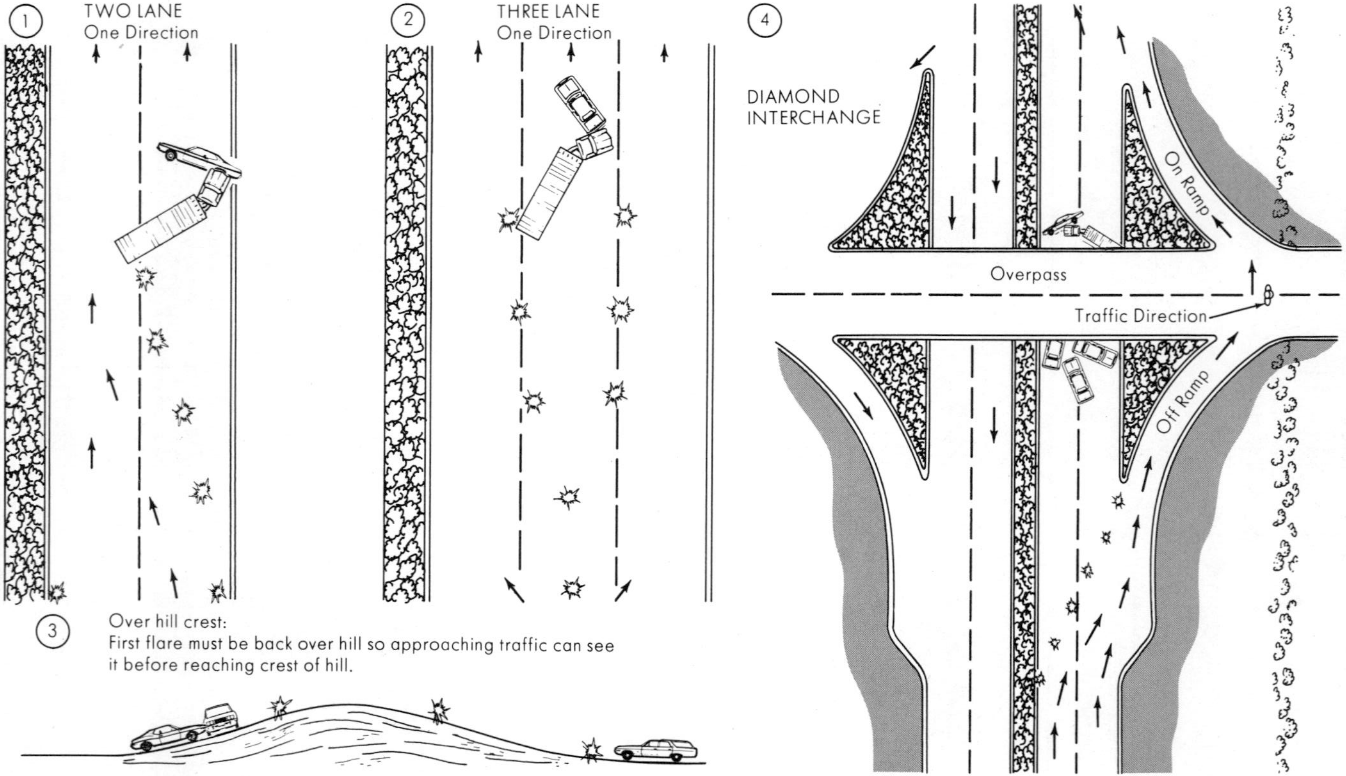

FIGURE 7—13
Freeway Flare Positions

considered in developing a traffic flare pattern at a scene. That is, speed should always be a consideration. The stopping distance chart below can be used to determine how far to start a flare pattern from the obstruction. After ascertaining the speed limit and prevailing speeds of traffic, the officer computes the total stopping distance by adding the thinking distance and braking distance found on the stopping distance chart. The computed distance is the distance the flare pattern should begin away from the obstruction.

Stopping Distance Chart

PASSENGER CARS ONLY

Speed	Thinking Distance	Braking Distance	Total Feet
25	27 feet	39 feet	66
35	38 feet	76 feet	114
45	49 feet	125 feet	174
55	60 feet	187 feet	247
65	71 feet	261 feet	332

Thinking Distance—distance traveled before brakes are applied while driver is reacting to danger.

Braking Distance—distance traveled after brakes have been applied.

Hazards Associated with Flares

The greatest danger associated with the use of flares occurs during the lighting process. Simple safety rules to be observed are:

1. Ascertain if there is a fire danger before lighting flares, such as leaking gasoline, butane, and other flammable substances.
2. Turn head and eyes away before striking flares because they have a tendency to pop when struck.
3. Keep flares away from your body by extending your arm because dripping molten material can cause serious burns.
4. Stand upwind from flares and hold them downward when using a flare to direct traffic because the toxic fumes being emitted can be harmful if breathed for prolonged periods.
5. Grasp the flare by the end away from the flame and gently tap the burning end on the pavement, then snuff out the flame when the flare is no longer needed.
6. When placing flares at an accident scene, always walk toward oncoming traffic when placing flares.

Techniques for Lighting Flares

1. Wax sealed cap type

    a. Pull black tab to break the seal and to expose the striking surface on the cap.
    b. Twist the cap off and remove to expose the igniting device of the flare.
    c. Point flare down and away.
    d. Hold cap with striking surface in the weak hand and flare in the strong hand.
    e. Strike the striking surface of the cap downward and away against the igniting device.

2. Plastic Cap Type

    a. Twist plastic cap off to expose igniting device.
    b. Remove small cover from cap to expose the striking surface of the cap.
    c. Continue the same steps as with the wax sealed type.

3. If on a grade, in high wind conditions, or if the movement of traffic will make the flares roll, use the cap from the flare.
    a. After lighting the wax sealed flare:
        (1) Stuff the loose end of the black tab into the cap, forming a loop with the tab. (This should be a reverse loop so that the black top is exposed.)
        (2) Place the cap on the rear of the flare (see fig. 7–14).
    b. After lighting the plastic cap type flare:
        (1) Place the plastic cap on the rear of the flare and the extended tip will prevent the flare from rolling (fig. 7–15).

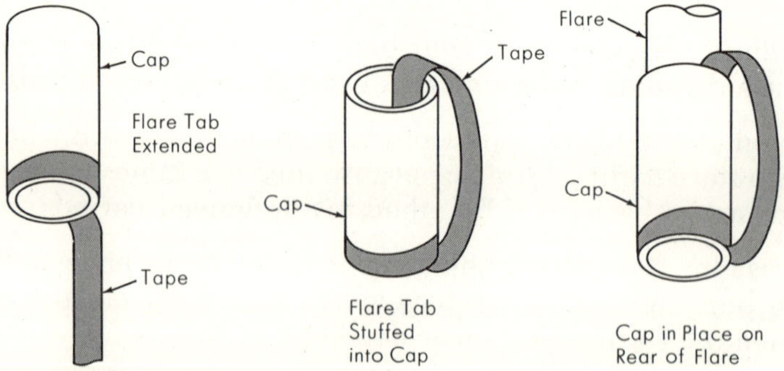

FIGURE 7–14
Wax Sealed Type Flare

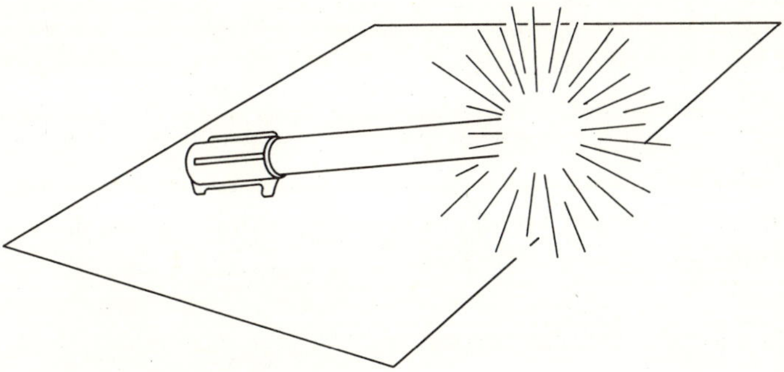

FIGURE 7—15
Plastic Cap Type Flare

Summary

Traffic direction, in this chapter, was approached on a practical basis rather than a theoretical one. Therefore, its content was geared more toward practical application. The importance of using uniform signals and gestures for starting, stopping, and turning traffic was stressed. The use of signalling aids such as the voice, whistle, and flashlight were explained along with their relative benefits and limitations.

Intersection control can be simple or complex depending upon the intersection type, location, and traffic therein. Important responsibilities associated with intersection control were covered and important traffic variables were outlined. The recommendation to follow traffic signal sequences at signalized intersections was given prime consideration as well as the positioning of the officer within an intersection.

Although accidents are not the only event requiring flare patterns to warn oncoming traffic, they are the most prevalent. Consideration was given to the importance of establishing priorities in order to handle emergencies according to their importance.

The proper placement and alignment of flares was explored. Aids for establishing flare distance requirements such as the stopping distance chart were offered for consideration as well as a method for computing thinking distance or reaction distance.

Proper utilization of traffic direction techniques and proper utilization of flare patterns will operate to facilitate the function of the traffic control administrator.

DISCUSSION QUESTIONS

1. Discuss the importance of uniform signals and gestures.

2. Discuss the reasons why traffic direction at signalized intersections should follow the normal signal sequence.

3. Discuss why the voice as a signalling aid is seldom used.

4. Discuss the importance of avoiding decision-making situations when laying flare patterns.

5. Discuss why speed should always be considered when developing a flare pattern.

8: Traffic Laws

Traffic laws evolved with increased use and complexity of automobiles and highway service at the turn of the century. Prior to the development of the automobile, people were accustomed to vehicles being drawn by animals. Maximum speeds were about six miles per hour. The relatively slow speed and animals which instinctively avoided contact with other animals and vehicles provided a relatively comfortable traffic scene. Since then, the traffic scene has changed rapidly. From about 1900 until the present time highway speeds, not necessarily legal speeds, have increased to over 100 miles per hour, and the animal with its instinct for self-preservation has been replaced as the source of power. The increased speed and the loss of an inherent protective system have contributed to the creation of a devastating traffic environment.

Early Traffic Laws

The capacity for greater speed developed faster than people's ability to contend with it. Thus, speed became the focus of controls and restrictions early. Nationally, during the early 1900s speed restrictions were among the first traffic laws enacted. The first state to have a statute regulating speed was Connecticut. In 1901, its speed was limited to twelve miles per hour in town and fifteen miles per hour outside of town.[1] Along with speed laws

---

1. Edward C. Fisher, *Vehicle Traffic Law*, ed. Robert L. Donigan (Evanston, Ill.: The Traffic Institute, Northwestern University, 1970), p. 39.

came the enactment of traffic laws which applied to vehicles meeting, passing, signalling and operating with specific safety equipment.

In contrast, speed was not among the first laws pertaining to the use of vehicles in California at the turn of the century. Those laws in existence stated that all vehicles meeting on a roadway must pass on the right and that one vehicle should give one-half of the roadway to the other vehicle. Later, around 1905, laws were enacted to regulate vehicle registration, drivers' licenses and speed. A seal, two inches in diameter, was issued to be displayed on each registered vehicle. A small badge was issued to every chauffer, operators for hire, who was required to have a driver's license. Speed limits of ten miles per hour were set for in-town travel and twenty miles per hour for out-town travel.

Need for Uniform Traffic Laws

With the mass production of automobiles by Henry Ford (1917–1918), the status of automobiles quickly moved from being a pleasure item to an item of necessity. Since that time the need for traffic laws has become progressively greater. To meet this need, traffic law legislation was provided by various governmental bodies at local and state levels. All too often the laws enacted were not representative of or did not have direct application to the traffic problem. At best, attempts to resolve the traffic problem through legislation were fragmented. Invariably, these fragmented approaches were not comprehensive enough to deal with the real issues. In addition, many of these laws lacked uniformity and were in conflict with each other geographically. Local laws conflicted with state laws, and state laws conflicted with other state laws. This situation created havoc for the motorist traversing geographical boundaries. Further complicating the situation was the lack of uniformity of traffic law enforcement.

Uniform Vehicle Code

It was soon recognized that uniform traffic laws with national application were sorely needed. What was not needed were laws so binding or rigidly structured that local problems could not be resolved. To deal with this need, a National Committee on Uniform Traffic Laws and Ordinances was formed. In 1926, after much hard work, the first issue of the Uniform Vehicle Code was published. Many states adopted this code as a guideline for state laws.

Two years later the Model Traffic Ordinance was published.[2]

The design of the Uniform Vehicle Code was to provide a comprehensive guide and standard for state motor vehicle and traffic laws. It reflected the need for uniform traffic laws throughout the United States. The concept of uniform laws did not mean that all laws must be the same in all aspects of the highway traffic environment. However, similar situations should be treated with a high degree of consistency with regard to the type of action contemplated.

The purpose of the Uniform Vehicle Code was to provide a guide for the establishment of reasonable and comprehensive traffic codes within each state. The philosophy of the National Committee on Uniform Traffic Laws and Ordinances can be found in their motto *Salus, Libertas, Lex*—"Safety with Freedom Through Law."[3] The idea is fostered that traffic legislation is not to impose unnecessary or unreasonable restrictions on highway traffic, but to insure that traffic shall flow smoothly, expeditiously, economically, and safely. Further, all highway users should have reasonable protection from being killed, injured, or frustrated by the improper driving practices of others.

Right to Regulate Traffic

Needless to say, the issues as to the right of states to regulate traffic was raised. In due time this issue was clarified and now the power of states to regulate the use of public highways is quite clear. The burden of legislating reasonable traffic laws and providing a traffic environment where the drivers have a reasonable expectation as to how to behave has become the responsibility of most states. Today, all states have reasonably comprehensive traffic codes.

State traffic codes patterned after the Uniform Vehicle Code commonly provide that, except as otherwise expressly provided, the provisions of their codes are applicable and uniform throughout the state and in all counties and municipalities therein. No local authority can nor shall enact or enforce any ordinance which is in conflict with the state code, unless it is expressly authorized. It has been generally accepted by most states that

2. National Committee on Uniform Traffic Laws and Ordinances, *Uniform Vehicle Code and Model Traffic Ordinance* (Alexandria, Va.: Michie Company, 1968), p. v.

3. Ibid., p. vii.

municipalities may legislate on the same subject so long as the municipal ordinance does not conflict with the state law, and if there is no conflict between them, both laws may stand. [4]

However, California is one of the few states limiting the authority of municipalities to enact laws. Local authorities cannot enact or enforce any ordinance on matters covered by the California Vehicle Code, unless expressly authorized therein. This means that ordinances substantially similar and those designed to clarify state laws are invalid. Local authorities can enact laws which are supplementary to or in addition to state laws. They have no power to legislate on matters which are not purely local in nature or on control of traffic which is not a local matter.[5] Similarly, statutes in Indiana and New Jersey prohibit enactment of ordinances covering offenses defined by state law. The Supreme Court of Georgia in *Hannah* v. *State*, 97 GA. App. 188, 102 S.E. 2d 6214 (1958) held that municipalities do not have the right to duplicate provisions of state laws. The rationale of this case was that ordinances are "local and special laws."

Criminal Law Implications

In order for traffic laws to become operative they must be consistent with the principle of criminal law. Criminal law criteria vary among states; however, the degree of differentiation is not that great. In most states violations or offenses take place when prohibited acts are committed or when required acts are omitted. In essence, a crime is committed when one or the other occurs.

Crime is defined as the commission or omission of an act for which there is a prescribed penalty. In most states no act is criminal unless expressly provided by statute. In support of this requirement, the Supreme Court of Nebraska in *Dutiel* v. *State*, 135 Neb. 811, 284 N.W. 321, 324 (1939), ruled that "no act is criminal unless the legislature has in express terms declared it to be so, and no person can be punished for any act or omission which is not made penal by the plain import of the written law." Yet there are some states which operate under the concept of "common law" which consists of doctrine that is not set down in any written statute or ordinance and depends merely upon time-honored usage.

---

4. Fisher, *Vehicle Traffic Law*, p. 126.
5. *Abbot* v. *City of Los Angeles*, 53 Cal. 2d (Adv) 679, 3 Cal. Rptr. 158, 349 P.2d 974 (1960).

Traffic law statutes consist of various elements (*corpus delicti*). These elements must be present in each circumstance before an offense is complete. When all the elements are present the offense is complete regardless of the state of mind of the violator. That is, no specific intent is required of the violator to break the law unless it is specifically expressed in the statute as an element of the offense.[6] Most traffic laws are formulated to make an act or omission unlawful regardless of whether there was intent to violate the law or to do wrong.

Classification of Offenses

Most states have codified provisions for the classification of crime for each violation of law. The most common classification of crimes are felonies and misdemeanors. Some states pattern their classification after federal statutes. Federal statutes identify three categories of crimes: (1) felony, (2) misdemeanor, and (3) petty offense. In the area of traffic laws, some states, along with California and New York, classify some traffic offenses as traffic infractions. In Pennsylvania violations of traffic laws and regulations relating to the rules of the road are designated as "summary offenses" and are not considered crimes.[7]

Generally, a felony is a crime which is punishable by death or imprisonment in a state prison, and every other crime or public offense is a misdemeanor except those offenses that are classified as infractions. A traffic infraction is not a crime, and the punishment imposed should not be deemed for any purpose a penal or criminal penalty or punishment. For example, traffic infractions in California are treated as follows: they are not punishable by imprisonment; violations require the payment of fines only; and persons charged with an infraction are not entitled to a trial by jury or the services of a public defender.[8]

Traffic Laws and Interpretations

During our research of traffic laws we found that those states whose laws are closely patterned after the guidelines of the Uniform Vehicle Code are relatively uniform in nature. Since a high degree of similarity of traffic laws exists between states, significant Uniform Vehicle Code sections will be covered in terms of

6. *State v. Sonoderleiter*, 251 Iowa 196, 99 N.W. 2d 393 (1959).
7. Pennsylvania, *Vehicle Code*, sec. 1202 (1958).
8. California, *Penal Code*, sec. 19c (1974–1975).

application and court interpretation. It is hoped that the scope of this approach will be broad enough to have widespread application. Because of obvious limitations it is suggested that respective state vehicle codes be used as a primary reference for local and state laws and as a supplement to this chapter.

The clarity of traffic laws is enhanced by having a clear understanding of terms commonly used. Major categories to be covered are: (1) components of the traffic environment, (2) components of the highway system, and (3) traffic environmental controls.

Components of the Traffic Environment

Components of the traffic environment are of various types. As used in the following context, they are those elements of the traffic environment which are subject to traffic regulations. These elements are identifiable with the term "traffic," and "traffic" is defined by section 1-177 of the Uniform Vehicle Code as:

> Pedestrian, ridden or herded animals, vehicles, streetcars and other conveyances either singly or together while using any highway for purpose of travel.

Within the traffic environment some traffic components are encountered more frequently than others. The ones which are most commonly encountered are vehicles, pedestrians, bicycles, and ridden animals. While using the highway for purposes of travel each is subject to traffic regulations.

Traffic regulations are designed to control traffic in order to provide a safe and economical traffic environment. Most traffic laws deal with the operation of "vehicles" upon the highways. The term vehicle is too broad and a more succinct definition is needed. The Uniform Vehicle Code, section 1-184, states that a vehicle is:

> Every device in, upon or by which any person or property is or may be transported or drawn upon a highway, excepting devices moved by human powers or used exclusively upon stationary rails or tracks.

The classification of a vehicle is modified with the addition of the term "motor"; thus, "motor vehicle" becomes an identifiable term. Also, many traffic laws define violations in terms of "motor vehicle" operation. The Uniform Vehicle Code, section 1-134, defines "motor vehicle" as:

> . . . every device which is self-propelled and every vehicle which is propelled by electric power obtained from overhead trolley wires, but not operated upon rails.

An essential part of the operation of a vehicle is the "driver." A driver is "every person who drives or is in actual physical control of a vehicle."[9] To drive a vehicle means to control its speed or direction while in motion. Thus, steering a car being pushed by another is "driving" it, since the person steering the car is controlling *its* direction.[10] In contrast, steering a towed car is not "driving" since the person steering cannot control the direction of the car.

The operation of a vehicle, which is an integral part of the driving of a vehicle, requires actual physical handling of the controls. One is operating a vehicle when starting the motor of a vehicle; steering a vehicle with the engine shut off while coasting down a sloping highway; or pushing and steering a disabled vehicle down a highway.[11] Being in physical control of a vehicle not only includes "driving" and "operating." It is more inclusive than either as was found in the Arizona case of *State* v. *Webb*, 78 Ariz. 8, 274 P. 2d 338 (1954). In this case "One found asleep in his car with head and arms resting on the steering wheel, the motor running and lights on, is in 'actual physical' control of the vehicle."

Another component of the traffic environment is the "pedestrian." A pedestrian is "any person afoot" as defined by section 1-143 of the Uniform Vehicle Code. The actions of pedestrians are regulated and their responsibilities according to section 11-501, Uniform Vehicle Code, are:

> (a) A pedestrian shall obey the instruction of any traffic device specifically applicable to him, unless otherwise directed by a police officer.
>
> (b) He is subject to traffic and pedestrian control signals.

The activities of "ridden or herded animals" within the traffic environment are provided for and limited by section 11-104, Uniform Vehicle Code. It states:

> Every person riding an animal or driving any animal-drawn vehicle upon a roadway shall be granted all the rights and shall be

9. *Uniform Vehicle Code*, sec. 1-114 (1968).
10. *Hester* v. *State*, 196 Tenn. 680, 270 S.W. 2d 321, 47 ALR 2d 568-570 (1954).
11. Fisher, *Vehicle Traffic Law*, p. 276.

> subject to all duties applicable to the driver of a vehicle . . . , except provisions . . . which by their very nature can have no application.

In modern-day traffic, the presence of bicyclists is ever increasing. Of the various devices called bicycles only those meeting the definition of "bicycle," as defined by statutes, are subject to traffic regulation. Section 1-105, Uniform Vehicle Code, defines "bicycle" as "every device propelled by human power upon which any person may ride, having two tandem wheels either of which is more than 14 inches in diameter." According to section 11-1202, Uniform Vehicle Code, "every person riding a bicycle upon a roadway shall be granted all the rights and shall be subject to all the duties applicable to the driver of a vehicle . . . except as to those provisions . . . which by their very nature can have no application."

The components of the traffic environment discussed above are not considered exhaustive. The list could be expanded to include variations of those mentioned. For example, motor vehicles involve motor trucks, motorcycles, school buses, and specially constructed vehicles. There are limitations imposed on each. The scope of this chapter is limited to broad categories with the intent of conveying some basic concepts and how they relate to the whole scheme of traffic control and investigation.

Components of the Highway System

As stated earlier, traffic laws are designed to control the operations of vehicles upon the highways in order to provide a safe and economical traffic environment. We will now look at some of the components of the highway system. These components are part of the arena where traffic activities are engaged. For clarity and understanding, again there is a need to define some terms. Inherent within some of the terms are the limits of the traffic environment. With the exception of a few traffic laws which cover off-highway operation of vehicles, the scope of most traffic laws are confined to the operation of vehicles upon the highways.

Alley. An "alley" as a component of the highway system imposes the same obligation on users as any other street or highway. That is, the usual rules of the road apply. By definition an alley is "a street or highway intended to provide access to the rear or side of lots or buildings in urban districts and not intended for the purpose of through vehicular traffic."[12] Some jurisdictions, by statutes, limit the width of alleys to twenty-five feet or less, exclude alleys intersecting other highways from forming "unmarked crosswalks," and extend alleys to unincorporated areas (communities outside of cities).[13]

Variations in the definition of "alley" have elicited a variety of court interpretations. In the case of *Irwin* v. *City of Manhattan Beach*, 51 Cal. Rptr. 881 (1966), it was held ". . . alley does not include 'street' as the lesser does not include the greater. Further an alley differs from a street in quality as well as quantity, in that 'it is a right of way . . . to serve a limited neighborhood for local convenience and not for general passage or travel as in the case of streets,' 64 C.J.S. Municipal Corporation, sec. 1653, p. 24."

The Uniform Vehicle Code, section 1-102, limits alleys to urban districts. As held in North Carolina in the case of *Faw* v. *Town of North Wilkesboro*, 253 N.C. 406, 117 S.E. 2d 14 (1960), "'alley' relates exclusively to a way in a town or city." This case implies that by definition limitations can be set as to where alleys can exist. Namely, they do not exist outside of towns or cities in North Carolina.

Crosswalk. Since highway users are required to yield the right of way to pedestrians using a "crosswalk," it is important to clarify highway conditions which impose such duties. There are two basic conditions regarding the existence of crosswalks. They are "unmarked crosswalks" and "marked crosswalks." The case of *Well* v. *Alderman*, 117 Ga. App. 724, 162 S.E. 2d 18-23 (1968), brought this out when it was ruled that "only two kinds of crosswalks are recognized in the law—a marked or painted one which may or may not be at an intersection, and an unmarked one which may be only at an intersection."

An "unmarked crosswalk" can be defined as "that part of a roadway at an intersection included within the connection of lateral lines of the sidewalks on opposite sides of the highway measured from the curbs or, in the absence of curbs, from the edges of the traversable roadway."[14] All states do not include the phrase ". . . connection of lateral lines of the sidewalk on opposite sides of the highway . . . ." However, those states using this phrase by definition imply that an "unmarked crosswalk" can exist only if a street's sidewalk is intersected and continues beyond such inter-

12. *Uniform Vehicle code*, sec. 1-102 (1968).
13. California, *Vehicle Code*, sec. 110 (1974).
14. *Uniform Vehicle Code*, sec. 1-111(a) (1968).

section. Therefore, "unmarked crosswalks" do not exist at "T" intersections at those locations where the sidewalks do not continue beyond the intersection (see fig. 8–1).

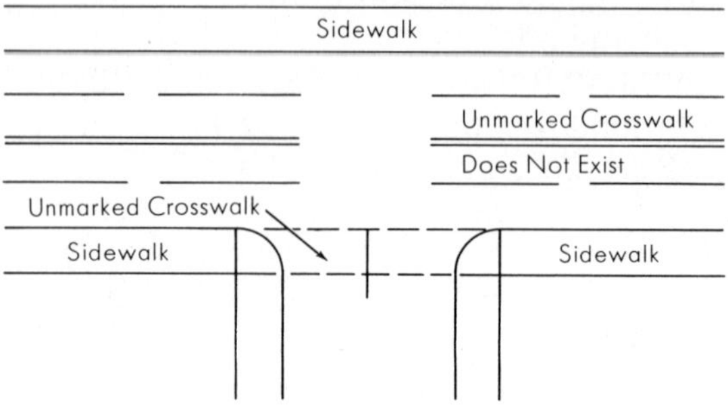

FIGURE 8—1
"T" Intersection

The Supreme Court of Georgia in *Griffin* v. *Odum*, 108 Ga. App. 572, 133 S.E. 2d 910-912 (1963), has held that "there is no 'unmarked crosswalk' where there is no sidewalk on the opposite side of the street with which 'lateral lines' may connect." This case was responding to the Georgia Vehicle Code section defining "crosswalk" in conformance with the Uniform Vehicle Code section on "crosswalk."

At least one state's statute excludes ". . . prolongation of such lines from an alley across a street" from the definition of an "unmarked crosswalk." Thus, as previously mentioned under "alley," "unmarked crosswalks," in that state, do not exist where alleys intersect with other streets or highways.[15]

The other crosswalk condition, the "marked crosswalk," is straightforward and clear. Its Uniform Vehicle Code definition, "any portion of a roadway at an intersection or elsewhere distinctly indicated for pedestrian crossing by lines or other markings on the surface," is self-explanatory.[16]

Highway. The Uniform Vehicle Code, rules of the road, refers exclusively to the operation of vehicles upon highways. State statutes, likewise providing for the commission or omission of acts upon highways, necessarily become one of its essential elements. The Uniform Vehicle Code, section 1-122, defines "highway" as "the entire width between the boundary lines of every way pub-

15. California, *Vehicle Code*, sec. 275 (1974).
16. *Uniform Vehicle Code*, sec. 1-111(b) (1968).

licly maintained when and part thereof is open to the use of the public for the purposes of vehicular travel" (see fig. 8–2). This definition also defines "street." "Street" and "highway" are synonymous and interchangeable terms. Those locations outside of these parameters, as a general rule, do not apply to the operation of vehicles. Streets and highways are not just for the use of vehicles only. As stated in *Armstrong* v. *Sengo*, 17 Cal. 2d 300 (1949), "streets and highways are maintained for the use of pedestrians as well as for vehicles."

Included within the boundaries of highways are shoulder or sidewalk areas of the road.[17] The shoulder is not required to be paved to be part of the highway. As held in a Virginia case, *Crouse* v. *Pugh*, 188 Va. 156, 49 S.E. 2d. 421, 4 ALR 2d 1242–1249 (1948), "the dirt shoulder of a highway is included therein, since it is built not only for the support of the paved surface but also for the protection and safety of highway users. It provides a place where users may drive off the hard surface when necessary, hence is a place 'open to the use of the public for vehicular travel.'"

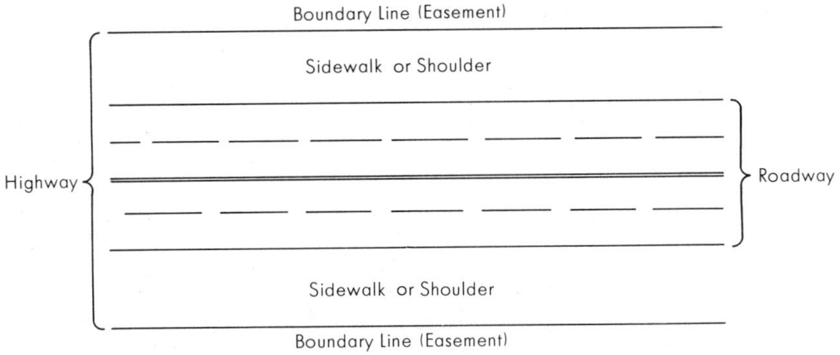

FIGURE 8–2
Highway

Intersection. Traffic laws prescribing the movement of vehicles at, near, or in an intersection exist in many jurisdictions. Whether such traffic regulations are applicable in a particular situation is frequently determined by what constitutes an intersection within the meaning of the law. The Uniform Vehicle Code, section 1-126, definition of "Intersection" is:

17. *Mecchi* v. *Lyon Van and Storage*, 38 Cal. 2d 474 (1940).

(a) The area embraced within the prolongation or connection of the lateral curb line, or, if none, then the lateral boundary lines of the roadways of two highways which join one another at, or approximately at, right angles, or the area within which vehicles traveling upon different highways joining at any other angle may come in conflict (see figs. 8–3 and 8–4).

(b) Where a highway includes two roadways 30 feet or more apart, then every crossing of each roadway of such divided highway by an intersecting highway shall be regarded as a separate intersection. In the event such intersecting highway also includes two roadways 30 feet or more apart, then every crossing of the two roadways of such highway shall be regarded as separate intersections (see figs. 8–5 and 8–6).

(c) The junction of an alley with a street or highway shall not constitute an intersection (see fig. 8–7).

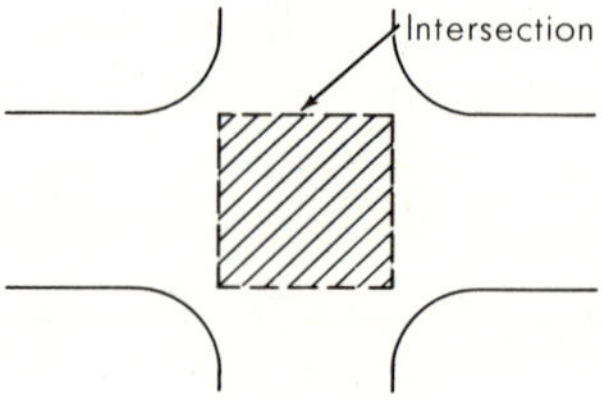

FIGURE 8–3

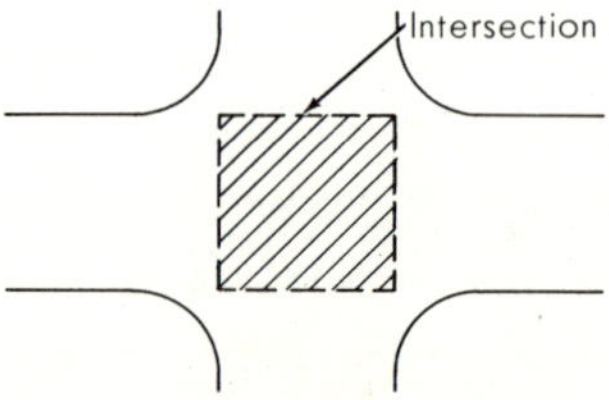

FIGURE 8–4

In *Connors* v. *Dobbs*, 77 Ohio App. 247, 32 Ohio Ops. 552, 66 N.E. 2d 546 (1942), it was said that "where two improved streets or highways intersect, the term 'intersection' means the improved area embraced within the lateral boundary lines of the two streets or highways, as improved, extended to their point of juncture with the lateral boundary lines of the intersecting street and highway, as improved." We find that this court's reasoning is consistent with the Uniform Vehicle Code definition as far as it goes.

In a number of cases an "intersection" has been construed under the applicable traffic regulation as the area in which vehicles traveling on the respective street may come into conflict.

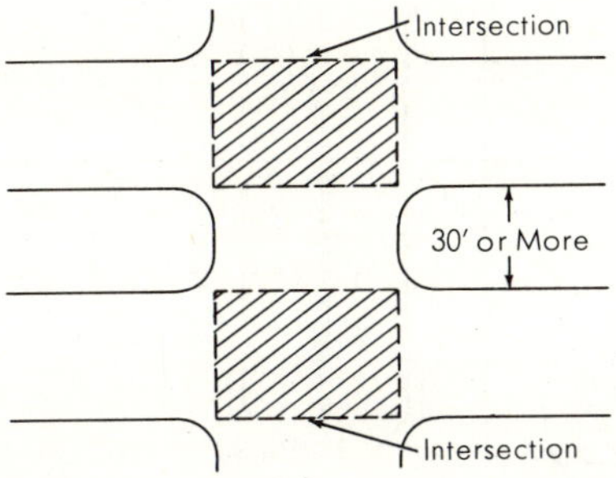

FIGURE 8–5

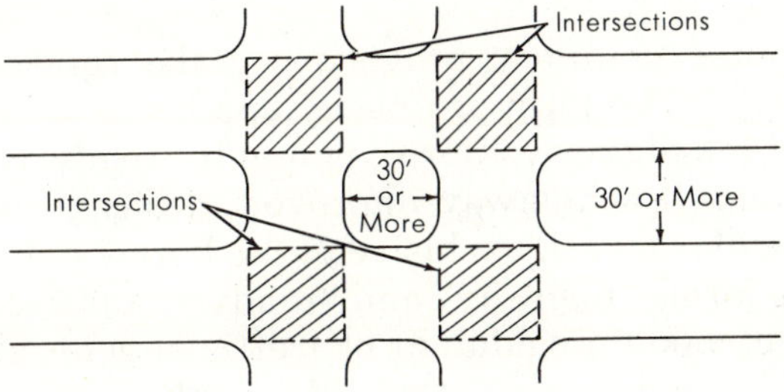

FIGURE 8–6

In the case of *Shedlock* v. *Marshall,* 186 Md. 218, 46 A2d 349 (1946), it was held that "an accident occurred within the intersection of two highways intersecting at acute angles when the cars traveling upon respective highways came into conflict." However, in the case of *Bell* v. *Huson,* 180 Cal. App. 2d 820, 4 Cal Rptr. 716 (1960), it was found that "a collision did not occur within the intersection when all the evidence indicating the point of impact was outside of the boundary of two highways intersection at approximately 55 degrees and outside of the path of the normal flow of traffic which must be considered in determining the area of the intersection."

Strict interpretation of the Uniform Vehicle Code definition of "intersection" seems to suggest that no "intersection" exists where intersecting streets or highways do not cross. This thought is derived from the emphasis placed on the prolongations of lateral boundary lines. However, in the case of *Commonwealth* v. *Cassidy,* 209 Mass. 24, 95 N.E. 214 (1911), it was brought out that "intersecting way shall mean any way which joins another at an angle, whether or not it crosses the other." Therefore, "a so-called 'T'-intersection at which one street meets with another at an angle without crossing the other is generally held to constitute an 'intersection' within the means of the application of traffic regu-

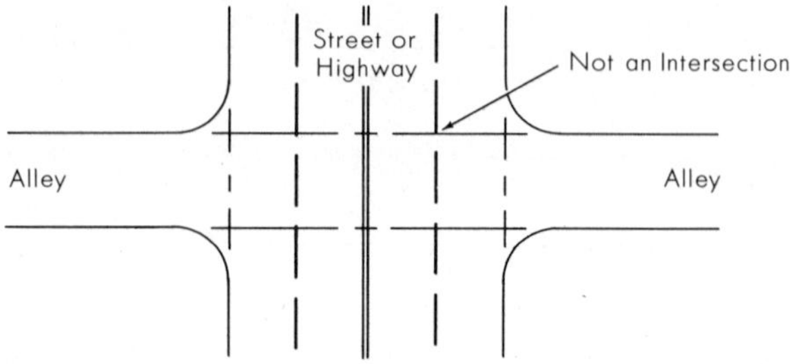

FIGURE 8—7

lations."[18] It should be noted that the same principles apply to the so-called "Y" intersections.

Roadway. Some statutory provisions make references to the term "roadway." The term ordinarily applies to the main travelled portion of the highway. More commonly "roadway" is defined as "that portion of a highway improved, designed or ordinarily used for vehicular travel, exclusive of the berm or shoulder" (see fig. 8—2). The terms "highway" and "roadway" as used in statutes are not synonymous. Regulation of traffic is generally in reference to the movement of vehicles on the roadway of a highway.[19]

During hours when designated parking lanes incorporated with the highway system are occupied, some have concluded that the "roadway" was exclusive of the parking lanes. However, in the case of *Shachunazarcan* v. *Widmer,* 159 Cal. App. 2d 180 (1958), it was held that "a city street is a 'roadway' from curb to curb, since it is open to travel in all of its parts." Generally, under the components of the highway system, we have established the environment within which traffic activities can be regulated. Now we will proceed to traffic environmental controls.

Traffic Environmental Controls

There are a host of traffic regulations covering most conceivable traffic situations in need of controls. A driver, in order to contribute to the safety of the traffic environment, must have some expectation as to what he or she is being held responsible for while using the highway system. To deal with this issue and as a limiting factor, Uniform Vehicle Code sections selected for discussion were chosen largely because of judgmental factors. That is, these sections require the exercise of reasonable judgment. In many cases the exercise of judgment has been lacking primarily due to lack of understanding of the traffic regulations.

---

18. *United States Fidelity and Guaranty Co.* v. *Duet,* La. App. 1, 177 S.O. 2d 302 (1965).

19. *Bates* v. *Legette,* 239 S.C. 25, 121 S.E. 2d 289, 292 (1961).

Traffic Control Devices. Traffic control devices such as traffic signs, signals, and markings are designed to regulate, guide and warn traffic. Their purpose is to minimize conflicts between highway users vying for the same space through the assignment of right of way. The Uniform Vehicle Code provides that:

> The driver of any vehicle shall obey the instructions of any official traffic control device applicable thereto placed in accordance with the provisions of this act, unless otherwise directed by a police officer, subject to the exception granted the driver of an authorized emergency vehicle in this act. [20]

One type of device provided is the traffic control signal legend which consists of sequential red, green, and yellow lights. Typically, red means stop, green means proceed and yellow warns of an impending red light. The greatest burden placed on a driver is the obedience to a red traffic signal. In terms of regulation it is provided that:

> Vehicular traffic facing a steady red signal alone shall stop at a clearly marked stop line, but if none, before entering the crosswalk on the near side of the intersection, or if none, then before entering the intersection and shall remain standing until an indication to proceed is shown. [21]

When one fails to stop for a red traffic signal often he or she raises the question of whether the signal was functioning properly at the time he or she proceeded through the intersection. A New York case dealing with this question, in response to an officer's testimony that the signal was functioning properly when he returned to check it, said "It is reasonable inference that a traffic device which was properly functioning a matter of minutes after a violation was also properly functioning at the time of the violation." [22]

Provisions of the Uniform Vehicle Code do not prohibit traffic from entering the intersection when the yellow or caution light is exhibited. The only requirement is that traffic not enter or be crossing the intersection when the red or "stop" signal is exhibited.[23] It must be shown the light changed to red prior to entry into the intersection in order to prove that a violation took place.

Following too closely. Many drivers are familar with the practice of following too closely behind a vehicle proceeding ahead.

20. *Uniform Vehicle Code*, sec. 11-201 (1968).
21. *Uniform Vehicle Code*, sec. 11-202 (c)1 (1968).
22. *People* v. *Elsessor*, 52 Misc. 2d 588, 276 N.Y.S. 2d 232-241 (1967).
23. *Atlanta Stone Work, Inc.* v. *Hollon*, 112 Ga. App. 862, 146 S.E. 2d 358, 361, 363 (1965).

Sometimes this is done deliberately and sometimes it is done inadvertently. In any event, it is an extremely dangerous practice which frequently results in personal injuries, death, or property damage.[24]

Statutes sometimes prescribe the distance to be maintained by a motor vehicle from the one ahead, some in general language such as a "reasonable and prudent" distance, and others in a specified number of feet. In a Maryland case it was held that the driver of a motor vehicle should not follow another vehicle more closely than was reasonable and prudent having due regard for the speed of such vehicles and traffic upon and the condition of the highway.[25] Typical of statutes requiring specific distances to be maintained is one noted in a Arizona case. It provided that, outside of a business or residence district, the driver of a motor truck should not follow another such truck within 100 feet but might overtake and pass another motor truck.[26]

In formulating a general rule respecting the distance to be maintained between motor vehicles, some courts have stated that the following motorist must remain behind a sufficient distance so as to be able to avoid a collision.[27] In other cases it has been held that this rule was applicable only so long as the driver ahead observed the rules of the road and that the following driver was not required to anticipate unusual contingencies.[28] In any event, the distance maintained according to the standard of reasonable care should be sufficient to avoid danger in the case of a sudden stop or decreased speed by the vehicle ahead under circumstances which are reasonably anticipated by the following driver.[29] Further, it is reasonable for a driver to assume that the car ahead will not stop instantaneously, but at the same time he or she cannot assume that the car ahead will always slow down and travel a short distance before coming to a complete stop.[30]

The Uniform Vehicle Code relative to following too closely provides that: "the driver of a motor vehicle shall not follow another vehicle more closely than is reasonable and prudent, having due regard for the speed of such vehicles and the traffic upon and the condition of the highway."[31] The similarity between this section and the language used in the Maryland case is obvious. Since the

24. 85 ALR 2d 613.
25. *Brehm* v. *Lorenz*, 206 Md. 500, 112 A. 2d 475 (1955).
26. *Southwestern Freight Lines* v. *Floyd*, 58 Ariz. 249, 119 P. 2d 120 (1944).
27. *Jackson* v. *Camp and Brown Produce Co.*, 92 Ga. App. 359, 88 S.E. 2d 540 (1955).
28. *Billington* v. *Schaal*, 42 Wash. 2d 878, 259 P. 2d 634 (1953).
29. *State* v. *Bush*, O.L.A. 161, 182 N.E. 2d 43-48 (Ohio 1962).
30. *Hibner* v. *Lindauer*, 18 Wis. 2d 451, 118 N.W. 2d 873-877 (1963).
31. *Uniform Vehicle Code*, sec. 11-310 (1968).

distance to be maintained is relative, it must be recognized that, in applying the statutory standard, the trier of fact must weigh a great many considerations, including speed, amount of traffic, road conditions and opportunities for clear vision.[32]

Right of way. Right of way rules are designed to standardize duties and obligations of highway users with respect to the establishment of priority rights between persons at a particular location in a particular situation. They furnish standards by which users may readily determine which has the right to proceed when both are about to use the same highway space. The primary purpose of the regulations is to minimize uncertainty and confusion at intersections and elsewhere and to make movement of traffic both practical and safe.[33] The Uniform Vehicle Code defines "right of way" as:

> The right of one vehicle or pedestrian to proceed in a lawful manner in preference to another vehicle or pedestrian approaching under such circumstances of direction, speed and proximity as to give rise to danger of collision unless one grants preference.[34]

Simply stated several courts, including California and New York, have said "Right of way' merely means a preference to one of two vehicles asserting the right of passage at the same place and at approximately the same time."[35] By definition it is implied that the question of right of way arises between two users of the highway only when there is danger of collision between them if both proceed on their respective ways without delay.[36]

An important qualification of right of way is that before anyone is entitled to it, the driver himself or herself must be proceeding in a lawful manner. One cannot acquire the right of way by violating the law. It was held in a Maine court that "right of way" rule presupposes approach of vehicles at lawful speed, otherwise no "right of way" is acquired.[37] In California it has been held that before a vehicle driver is entitled to right of way, he or she must be operating a vehicle within the law and not in violation thereof.[38]

The Uniform Vehicle Code covers various conditions in which right of way rules apply. Those selected for discussion are:

---

32. *Phillips* v. *Haring*, 262 Wis. 174, 180, 54 N.W. 2d 200-203 (1952).
33. *Martin* v. *Stevens*, 121 Utah 484, 243 P. 2d 747-750 (1952).
34. *Uniform Vehicle Code*, sec. 1-156 (1968).
35. *Cowan* v. *Market St. Ry.*, 8 Cal. App. 2d 642, 47 P. 2d 752-754 (1935).
36. *State* v. *McLachan*, 36 Cal. App. 2d Supp. 754 (1939).
37. *Ricker* v. *Morin Brick Co.*, 22 A. 2d 536-540 (Maine 1966).
38. *Carley* v. *Zeigler*, 156 Cal. App. 2d 643 (1958).

section 11-401, vehicle approaching or entering an intersection; section 11-402, vehicle turning left; section 11-403, vehicle entering a stop or yield intersection; and section 11-502, pedestrians' right of way in crosswalks.

Section 11-401 Vehicle Approaching or Entering Intersection

(a) When two vehicles approach or enter an intersection from different highways at approximately the same time, the driver of the vehicle on the left shall yield the right of way to the vehicle on the right.

(b) The right of way rule declared in paragraph (a) is modified at through highways and otherwise as stated in this chapter.

Paragraph (b) implies that paragraph (a) is designed to regulate traffic at uncontrolled intersections and on this basis we will proceed. Refer to the section on "intersection" for clarification as to what a street or highway intersection is within the traffic rule.

When two vehicles approach an uncontrolled intersection at approximately the same time, the vehicle on the left must yield the right of way to the vehicle on the right. The two vehicles must not only enter the intersection at the same time but also must by necessity have approached the intersection at approximately the same time.[39] The words "at approximately the same time" do not mean at precisely the same instant. In order for the vehicle on the left to secure the right of way it must enter the intersection at a sufficient interval of time ahead of a vehicle approaching from the right, so that it may be said that vehicle did not enter at approximately the same time. Such an interval must be of appreciable duration.[40]

If a driver is so far in advance of a vehicle on his or her right and has reasonable belief that he or she can clear an intersection without danger of a collision, the driver on the left may proceed into the intersection. He or she must not do so where the margin of safety is so narrow that a reasonably prudent person would not test an obvious danger.[41]

Section 11-402 Vehicle Turning Left

The driver of a vehicle intending to turn left within an intersection or into an alley, private road or driveway shall yield the

39. *Smith* v. *Murphy*, 48 Tenn. App. 299, 346 S.W. 2d 276-278 (1960).
40. *Brower* v. *Stolz*, 121 N.W. 2d 624-627 (N.D. 1963).
41. *Pugh* v. *Ludwig*, 409 Pa. 577, 186 A. 2d 911 (1963).

right of way to any vehicle approaching from the opposite direction which is within the intersection or so close thereto as to constitute an immediate hazard.

Responsibility is placed upon the driver of a car desiring to make a left turn to ascertain before attempting to do so that such maneuver can be made safely and without danger or undue delay to overtaking or oncoming traffic. He or she must refrain from making a left turn unless the way is clear. [42]

Generally, "immediate hazard" are words which are easily understood and require no special definitions. Their meaning is similar to "imminent peril," "timely warning," and "reasonable warning." As far as the application of law is concerned it has no precise set of measurements and must be judged on the basis of common sense in light of existing circumstances. [43]

Some drivers are of the opinion that if they are first in the intersection (controlled intersections) they have the right of way. To the contrary it has been found that where vehicles proceeding in opposite directions are stopped momentarily for a traffic control signal, the driver intending to turn left has no right to make a quick left turn across the path of the other, and being first in the intersection does not give him or her the right of way. A driver must give a plainly visible signal of his or her intention to turn and must ascertain if the turn can be made safely. [44]

Section 11-403 (b) Vehicle Entering Stop or Yield Intersection

Except when directed to proceed by a police officer or traffic-control signal, every driver of a vehicle approaching a stop intersection indicated by a stop sign shall stop at a clearly marked stop line, but if none, before entering the crosswalk on the near side of the intersection, or, if none, then at the point nearest the intersecting roadway where the driver has a view of approaching traffic on the intersecting roadway before entering the intersection. After having stopped, the driver shall yield the right of way to any vehicle which has entered the intersection from another highway or which is approaching so closely on said highway as to constitute an immediate hazard during the time when such driver is moving across or within the intersection.

The failure to yield right of way at a street controlled by a stop sign and failure to stop at a stop sign are separate offenses. One

---

42. *Desormeaux* v. *Continental Ins. Co.*, 153 So. 2d 128-138 (La. App. 1963).
43. *Lafferty* v. *Wattle*, 349 S.W. 2d 519-529 (Mo. App. 1961).
44. *Wiggins* v. *Ponder*, 259 N.C. 277, 130 S.E. 2d 402-404 (1963).

may be guilty of failing to come to a complete stop even though he or she yielded to oncoming traffic. Conversely, even though a driver came to a complete stop at a street controlled by a stop sign he or she may be guilty of failing to yield the right of way to traffic approaching so closely as to constitute an immediate hazard.[45]

If a reasonably prudent person does not believe he or she can pass through an intersection with safety, and a vehicle is approaching so closely as to threaten safety, mere stopping is not sufficient. He or she must remain stopped and shall not proceed until such movement can be made with safety.[46] If a vehicle is approaching on a highway close enough, considering the rate of speed, so that it would be reasonable to assume that a collision would occur if another vehicle stopped at the entrance of the intersection moved into its path, the stopped vehicle should remain stopped until it is safe to enter the intersection.[47]

The duty imposed upon a driver at a yield sign is less than the absolute duty to stop at a stop sign. The duty to stop at a stop sign is absolute, followed by a duty of lookout, including a calculation of interference with the right of way of other vehicles. The duty to stop at a yield sign arises after the required efficient lookout, including calculations.[48]

Section 11-502 Pedestrians' Right of Way in Crosswalks

(a) When traffic-control signals are not in place or not in operation the driver of a vehicle shall yield the right of way, slowing down or stopping if need be to so yield, to a pedestrian crossing the roadway within a crosswalk when the pedestrian is upon the half of the roadway upon which the vehicle is traveling, or when the pedestrian is approaching so closely from the opposite half of the roadway as to be in danger.

(b) No pedestrian shall suddenly leave a curb or other place of safety and walk or run into the path of a vehicle which is so close that it is impossible for the driver to yield.

A driver has the greater burden of careful lookout and extreme caution at locations provided for the passage of pedestrians, for it is the driver who has the power, because of the dangerous instrument which he or she controls, to reduce the status of a pedes-

45. *Parsekian* v. *Cresse*, 75 N.J. Super, 405, 183 A. 2d 426-429 (1962).
46. *Herman* v. *Wohler*, 419 P. 2d 45-47 (Ore. 1966).
47. *Magun* v. *Bemis*, 17 Wis. 2d 192, 116 N.W. 2d 129-133 (1962).
48. *Sailing* v. *Wallestad*, 32 Wis. 2d 435, 145 N.W. 2d 725-728 (1966).

trian to that of a statistic in a split second. The pedestrian poses no such threat. Because of this burden of response, a driver is entitled to adequate notice of the existence of a street condition which imposes upon him or her this special duty. A "marked crosswalk" normally would be clearly identifiable on the street by distinct and adequate markings.[49] Refer to the section on "crosswalk" for details of the two street conditions of crosswalks.

It may not be commonly known that stop signs do not apply to pedestrians. Stop signs are posted at intersections to control vehicular traffic only.[50] This does not mean that it is open season on drivers at intersections controlled by stop signs. Pedestrians are prohibited from suddenly stepping from the curb into the path of a vehicle which is so close that it is difficult for a driver to yield. A pedestrian is not entitled to the right of way even though he or she is crossing within a crosswalk at an intersection. Regulation patterned after the Uniform Vehicle Code, section 11-502(b), applies when a pedestrian unexpectedly leaves a place of safety at a time when a vehicle is so close it is virtually impossible to stop, but it does not apply when pedestrians are well into the intersection.[51]

Speed. Two major schools of thought have developed relative to speed restrictions. One group insists speed restrictions should be flexible, depending upon conditions and circumstances, leaving a wide area of determination for the prudence and judgment of the driver. From this school of thought the "basic speed law" concept evolved.[52]

The other school of thought insists the "basic speed law" with greater restrictions leaves too much to the judgment of drivers, many of whom cannot be relied upon to make proper decisions. Also, it contends such laws are virtually unenforceable since a question arises in every case, not merely as to the rate of speed, but almost exclusively as to the conditions and what was reasonable under the circumstances. Variances in physical and mental abilities of drivers along with limited enforceability of the basic speed law makes it essential to provide firm "absolute" limitation which drivers may violate at their peril. From this school of thought the "maximum speed law" concept evolved.[53]

Some states have a combination of the two speed laws. For example: (1) basic speed law with prima facie limits, and (2) an

49. *White* v. *McAllister*, 443 S.W. 2d 541-542 (1969).
50. *Folck* v. *Anthony*, 228 Md. 73, 178 A. 2d 413 (1962).
51. *Spann* v. *Ballesty*, 81 Cal. Rptr. 229 (1969).
52. Fisher, *Vehicle Traffic Law*, p. 223.
53. Ibid., p. 224.

overall (absolute) maximum beyond which it is unlawful to drivers under any circumstances. Provisions usually provide for reasonable and prudent speed, with *prima facie* limits in designated speed zones, and absolute maximum on the open highways. Consistent with uniformity the Uniform Vehicle Code provides for both restrictions.

> Section 11-801 Basic Rule
>
> No person shall drive a vehicle at a speed greater than is reasonable and prudent under the conditions and having regard to the actual and potential hazards then existing. Consistent with the foregoing, every person shall drive at a safe and appropriate speed when approaching and crossing an intersection or railroad grade crossing, when approaching and going around a curve, when approaching a hill crest, when traveling upon any narrow or winding roadway, and when special hazards exist with respect to pedestrians or other traffic or by reason of weather or highway conditions.
>
> Section 11-801.1 Maximum Limits
>
> Except when a special hazard exists that requires lower speed for compliance with section 11-801, the limits hereinafter specified or established as hereinafter authorized shall be maximum lawful speed, and no person shall drive a vehicle at a speed in excess of such maximum limits.

Practically all of the states embody the essentials of section 11-801 in their motor vehicle codes. It is the rule of safety stated in the form of a criminal law and applies regardless of the presence or absence of stated speed limits either absolute or prima facie.

What is meant by *prima facie* speed? *Prima facie* limits merely implement the basic speed law and serve as an aid in its enforcement. A *prima facie* case is complete when it is shown that the accused was exceeding the stated speed limit, that is, the law furnishes a presumption that the accused speed was greater than was reasonable and proper under the conditions. The degree of excess of speed over the stated speed limit is an important factor to be considered by the trier in determining whether, under all the circumstances, a motor vehicle has been operated at a speed greater than was reasonable.[54]

During prosecution, all the circumstances enumerated must be taken into consideration, but it is not required that all of them be other than reasonable at the time of the alleged violation. It is for

54. *State* v. *Simon*, 2 Conn. Cir. 642, 204 A. 2d 320-321 (1964).

the trier to determine under all of the circumstances, some of which may be favorable to the accused, whether or not the speed was greater than was reasonable at the time. [55]

Absolute maximum speed limits are arrived at through the use of sound engineering practices. They, in fact, represent reasonable and prudent limits and the speeds in excess thereof are fairly certain to be dangerous. Generally, maximum speed limits are based upon engineering and traffic surveys on existing highway segments, or upon the basis of approaching design standards and project traffic volumes in the case of newly constructed highway segments. [56] When maximum limits are set, they set a level which would facilitate the orderly movement of vehicular traffic and which would be reasonable and safe. Occasionally, other factors influence the establishment of maximum limits, such as the imposition of the 55 mph maximum speed limit for the purpose of energy conservation. Speeds in excess of this maximum limt are not necessarily unsafe as measured by reasonable standards of traffic safety.

Turning From Direct Course. Multilane highways have greatly promoted safe travel but at the same time their use has created some problems not associated with the two lanes of traffic proceeding in opposite directions. One problem is the movement of vehicles across right or outside lanes of a multilane highway to leave the road. Statutes in many jurisdictions require that right turns be made from the right lane of travel, or as closely thereto as practicable. [57] The Uniform Vehicle Code, section 11-601, covers these movements similarly. A driver who makes a right turn "wide" so as to permit a car to maneuver between him or her and the right curb or edge is in violation of such statutes. [58] This positioning criterion applies not only at intersections but also includes turning into private roads, driveways, or elsewhere. [59] When making a left turn a driver must exercise reasonable care under existing conditions to ascertain that such movement can be made with safety. It does not mean he or she has to refrain from making a left turn unless the circumstances are absolutely free from danger. Infallibility is not required. [60] However, the responsibility is placed upon the driver desiring to make a left turn to ascertain before attempting to do so that such movement can be made safely and

---

55. Ibid.
56. California, *Vehicle Code*, sec. 22356 (1974).
57. 7 ALR 3d 282.
58. *Luna* v. *Tecson*, 227 Cal. App. 2d 655, 39 Cal. Rptr. 24-28 (1964).
59. *Crouch* v. *Nicholson*, 116 Ga. App. 12, 156 S.E. 2d 384-387 (1967).
60. *Clarke* v. *Holman*, 274 N.C. 425, 163 S.E. 2d 783-786 (1968).

without undue delay to overtaking or oncoming traffic. He or she must refrain from turning left unless the way is clear. [61]

Before turning from a direct course upon a highway, the driver is responsible for seeing that such movement can be made in safety; that whenever the operation of any other vehicle may be affected by such movement, he or she must give a signal plainly visible to the drivers of such other vehicles of his or her intention to make the turn. The reponsibility for ascertaining that a left turn across a highway can be made safely is placed upon the driver attempting the movement. This rule applies to vehicles changing from one lane to another on multilane highways. [62]

Summary

Traffic laws have but one legitimate purpose, to establish standard rules of procedure to be followed by drivers in the operation of their vehicles so as to promote the safe and orderly flow of traffic. Early development of traffic laws brought about speed restrictions because automobile speed capacity increased faster than people's ability to cope with it. As the automobile increased in number and began to span greater distances, the impact of conflicting traffic regulation became great. Lack of standardization contributed to the confusion of drivers traversing geographical boundaries. The legitimate purpose of traffic laws during this period was not being fulfilled.

The need for uniform traffic laws nationally became evident. A National Committee on Uniform Traffic Laws and Ordinances was formed, and it was responsible for the publication of the Uniform Vehicle Code and Model Traffic Ordinance in 1926 and 1928 respectively. The acceptance of these publications as guidelines has been almost universal.

The evolution of traffic laws went through the processes of conforming to criminal law principals. They were defined as crimes and noncrimes of various classifications. They were classified as felonies, misdemeanors, petty offenses, infractions, and summary offenses or some combination thereof depending on the locale.

The application of traffic laws has not been without its difficulties. Over a period of time application of law has been reviewed by the courts and general rules and interpretations have been established. In this chapter an exploration of select Uniform

61. *Roy* v. *Edmonson*, 221 So. 2d 583-585 (La. App. 1969).

62. *Barker* v. *Thompson-Hayward Chemical Co.*, 152 So. 2d 649-651 (La. App. 1963).

Vehicle Code sections was made in terms of the types of highway traffic interaction within the traffic environment under defined constraints.

A study of court decisions of the period following the advent of the motor as a dynamic factor in our way of life revealed a gradual awakening of the judiciary to the true purpose of traffic laws as being preventive in nature. This is of great significance since it is in the courts that the law enforcement efforts culminate. Recognition that traffic laws are designed to prevent occurrence of a collision has been of inestimable value in the promotion of traffic safety through enforcement of laws enacted for that purpose.

DISCUSSION QUESTIONS

1. Identify and discuss the major problems confronted by the developers of early traffic laws.

2. Discuss the pros and cons of adopting uniform traffic laws for local traffic regulation.

3. Discuss the rationale for designating certain traffic laws as summary offenses or infractions.

4. Discuss why it is important to define elements of the traffic environment.

5. Discuss why it is important to define components of the highway system.

6. Discuss why it is sometimes difficult to apply statutory requirements of traffic laws.

9: Accident Investigation

Today, as in the past and as will be in the future, accident investigation will be a constant challenge to law enforcement agencies throughout the United States. Within our modern, contemporary industrial society the automobile has become an integrated, economic necessity. Without it our economy would collapse because neither people nor materials would be capable of movement sufficient to sustain our present way of life.

When physical objects are put into motion, there must be some inherent controls placed upon these objects or there surely will be conflict. The degree of conflict depends upon the space available for movement, the controls placed, the number and the size of the objects in motion. The objects we make reference to are motor vehicles, and the controls are a combination of drivers and the statutory vehicle laws (that govern our behavior) which affect the very operation of these objects. While chemists can effect compounds of materials which combine without occupying additional space, drivers of motor vehicles cannot so blend their conveyances without causing damage, injury, or death. Efforts are continually being made to enlarge the space available to motor vehicle and pedestrian traffic. As more streets are widened, freeways, expressways, toll roads, turnpikes, and the Interstate highway system is completed, so is there an increase in motor vehicle traffic. Space is not only something that we may regard as an intangible passage of openness, but it is also something that may be occupied by various objects of different size and shape. Therefore, every citizen should know that a combination of number and size of motor vehicles will likewise increase space. Thus, conflict is inevitable.

Education, Engineering, and Enforcement

Traffic crashes continue to cause damage and injury so extensive that all citizens are affected either directly or indirectly. Based on past experience, it is a foregone conclusion that traffic collisions cannot be eliminated, but rather they can be effectively reduced to an acceptable minimum. The problem is attacked by utilizing the techniques of *education, engineering,* and *enforcement.* Through education public interest is stimulated in order that all persons not only understand the extent of the problem, but they also become aware of the skills and attitudes necessary to reduce collisions. Properly engineered highways and vehicles reduce both the cause and effects of traffic collisions. The enforcement of governmental regulations aimed toward the reduction of traffic conflict, accidents, and their resulting damage and injury both guide and direct highway users in order to solve our society's transportation problem. Forceful implementation of each of these three measures will produce a three-pronged attack against the traffic problem within every community of our country, and which, if effectively done, will produce a more orderly system of mobility in our society. The real importance of accident investigation is to determine the who, what, when, where, why, and how these problems were created, and how they may be resolved so as to lessen the opportunity for reoccurrence. It is also of great importance to the police in relationship to other police activities, and likewise, it is important as a service to the community. Without a proficient and professional accident investigation program, the entire traffic supervision program will fail or be seriously weakened.

While fatal and critical injury accidents are usually investigated by law enforcement officers dispatched to the scene by police radio, less serious property-damage accidents may not require the same detailed investigation. A growing number of police agencies are conserving manpower by sampling less-serious accidents rather than attempting to investigate all of these property-damage-only or minor personal injury accidents.[1]

Information essential for effective planning of the "Three E's" is obtained mainly through accident investigation. Reports of acci-

1. Paul B. Weston, *The Police Traffic Control Function* (Springfield, Ill.: Charles C. Thomas, 1969), p. 164.

dent investigations provide one of the few sources of unbiased information available. These investigation reports perform a great public service. Accident investigation is, therefore, the cornerstone of the foundation of any system or program of traffic supervision. Without accurate information upon which to base plans, this function cannot be progressive or effective.

If planning to prevent accidents is to be valid and productive, it must be based upon consideration of accident causes. To achieve this, there must be careful investigation, together with complete and accurate reporting, to determine cause. The question then arises, which accidents should be investigated and reported? Fatal accidents alone, fatal and injury only, or to develop the complete picture—should all collisions be investigated?

The number of damage or injury accidents which occur for each fatality is unknown. Their incidence in cities, with slower speeds, is greater in proportion to the fatal accidents occurring in rural areas where higher speeds are attained. If planning were based on fatal accidents alone, only about 20 to 25 percent of the total reported accidents would be considered. This is but a small percentage of the total. If planning is expanded to include injury accidents, this base is expanded to about 55 percent of the collisions. However, 40 percent of the known crashes would still not be considered; thus, planning would still be based on only a little over half of the total accident occurrences.

The investigation and reporting of all known accidents is suggested for two reasons—adequate planning and good public relations. No matter how minor an accident may appear to one not directly affected, it is an important matter to those people actually involved. They look to the police for help. Although the police are primarily concerned with the criminal aspects of the investigation, parties concerned expect (and are entitled to) an impartial evaluation of the circumstances in order that the rights of all may be protected. The investigator should not attempt to establish civil liability or fix the blame, but he or she should impartially record the facts so that in any future civil action an unbiased, intelligent, and complete factual record will be available.

However, the time necessarily consumed by investigation of all accidents known to a department may not be available. There is validity in the argument that prosecution of the violator who has caused an accident is too late. It is proper to consider that the officer's presence on visible patrol, which deters other potential accident-causing violators, is time better spent than in the investigation of minor property damage collisions—especially those in which no violation is immediately discernible.

Instead of basing investigations upon the severity of accidents, the professional approach would appear to be consideration of the incidents in terms of possible violations—an enforcement viewpoint. Thus, a one-car fatality in which the victim was alone in his or her car might take a minimum of investigation (in spite of its grave consequences), as contrasted to a property-damage accident involving violations by both drivers. As an enforcement officer, the investigator must not lose sight of his or her objective. Obviously, the same amount of time cannot be devoted to all minor accidents because there are too many of them. Evaluation of the occurrence in terms of possible violation appears to be a firm basis to use as a guide to investigative effort.

The accident investigator should keep in mind throughout his or her investigation that the ultimate test of the quality of his or her work will be the convincing force it carries in the courtroom. Prosecution after the accident is one of the most equitable types of enforcement. Much time is spent in apprehending and citing drivers whose violations might cause an accident. If these drivers are prosecuted, it follows that in all fairness prosecution should be instituted against those whose violations have caused accidents. By pointing out the relationship between collisions and violations, the public may be brought to realize the need for strict enforcement of all traffic laws as a preventive measure. The entire enforcement program will be bolstered by public support as the realization grows that if traffic violations were not combated by enforcement, a terrific increase in highway casualties would surely result. The philosophy of "haven't they suffered enough already" is no valid reason for failing to prosecute. The fleeing felon who is seriously wounded during an escape attempt undoubtedly suffers from his or her wound. Society, however, spares no effort to keep him or her alive in order that he or she may stand trial for his or her offense. Why should personal suffering, no matter how great, limit prosecution of the traffic violator?

This type of prosecution is readily understood by the general public. People who witness, or are involved in, an accident where there is a violation expect the police to do something about it. They are perplexed if nothing is done. Prosecution of violators is an objective lesson to the general public, acts as a strong deterrent, and lessens opportunities for further deaths, injuries, and economic losses. It presents the officer to the public as someone who is doing something tangible for them. It engenders a feeling of protection which, in turn, creates a desirable relationship. To the violator it points out that all traffic laws must be obeyed.

In contrast to the crime rates, which steadily increase, the accident rate not only can be but also has been lowered. In the face of increased travel, America's mileage death rate (known as M.D.R.—deaths per 100 million miles traveled) has dropped. In a 1972 study, the M.D.R. was 7.7 deaths per 100 million miles traveled, during the first year of the study. The following year the rate was 5.3, and in 1974 it was down to 4.5. During this same time span, another study was being conducted on the population death rate (deaths per 100,000 population) which also showed signs of reduction. In 1972 it was 29.1, but by the last year of the study it had been lowered to 24.5. While it is true that the number of fatalities attributed to motor vehicle collisions has risen, it can be seen that the attack against the problem has been successful. The statistics prove that the chance of today's citizen being killed in an accident has been significantly reduced.

Accident investigation is not only important but also is interesting. The investigator deals with human problems, frailties, and suffering. In turn, he or she renders a service to humanity. His or her immediate action may directly save one life and indirectly many lives. Each investigation presents a new challenge, calling for the utmost in skill, knowledge, and ability.

Because of lack of motive, accident investigation is different from other types of police investigations. If the occurrence was planned, it would no longer be an accident. Accidents happen unexpectedly, without design, or by chance. The element of surprise is a part of any traffic accident.

Successful accident investigation requires considerable basic background knowledge. To know what questions to ask and what to look for the investigator must have some fundamental ideas about accidents and their causes. To avoid wasting time and making mistakes, especially during the urgent activities at the scene of an accident, the investigator needs to plan what he or she is going to do and to continually revise the plan as he or she proceeds.[2]

Traffic Collisions

During recent years, scientists have turned to the word "crash" in preference to the term "accident" in reference to traffic collisions. Their belief is that traffic collisions are *caused occurrences* rather than incidents involving chance alone. In almost every traffic crash there has been human failure—someone has made a

2. J. Stannard Baker, *Traffic Accident Investigator's Manual for Police* (Evanston, Ill.: The Traffic Institute, Northwestern University, 1966), p. 7.

mistake by neglecting to obey the dictates of either traffic laws or defensive driving. For police investigative purposes, however, both terms may be used interchangably since the officer's objective is to determine collision cause. Accident investigation offers a challenge to the investigator and gives him or her an unlimited opportunity to use his or her initiative, skill, and experience to uncover the factors which caused the crash.

Accident investigators are given the opportunity to perform a public service. Not only are the results of investigations utilized in research, but officers also render immediate assistance to those involved in the collision. They become the protectors of the suffering. The public views with favor actions to expedite traffic and assist victims of accidents. Such positive actions are necessary in order to elicit public support of enforcement organizations.

Many major arrests will follow accident investigations. These include arrests for misdemeanor and felony drunk driving, manslaughter, hit and run, reckless driving, auto theft, and other crimes. Obviously, a great amount of care must be used in case preparation. Essential evidence must be recognized, identified, and properly preserved. The investigator must be careful to make a thorough investigation, prepare his or her case meticulously, and present it well in court. The officer whose work is inadequate will fall short of accomplishing his or her duty to the people whose rights he or she is sworn to defend, protect and uphold.

In civil cases too, courts have come to rely upon the investigating officer for unbiased and informed testimony. The traffic accident investigator has acquired an enviable reputation in this connection, and judges and attorneys have come to give great weight to the officer's testimony. The investigator should always be conscious of this, and he or she must realize that a poor investigation or presentation of evidence will destroy that which has taken years to build.

In order for the officer to be able to relate the events of the crash, through results of a thorough investigation, to the court and/or other government and civilian agencies, it is necessary that he or she have a working knowledge of some of the causitive factors of these same accidents.

First we must put into some perspective how the puzzle is to be assembled. The operation of an automobile is like a puzzle, where all the pieces must fit properly into place in order that we may complete our destination. There must be consideration as to the route we are to take, the operation of the motor vehicle we are to drive and at the same time the alternatives available should something cause us to take evasive action when a hazard appears. One must, of course, be able to recognize a hazard. This calls for a certain amount of driving strategy, or evasive action of an operational nature. There are four steps in one's driving that may be a part of the crash potential. First, *recognize* the situation that may require some type of action. Second, *anticipate* the potential of an emergency situation. Third, *decide* what action is to be taken and also what *alternatives* are available. Finally, there must be *performance* of what has been decided or the alternative chosen.[3]

At this point in time the investigator may be able to assemble the pieces of the collision puzzle, which in turn will give the essential *how* of the crash, but they do not include the *why* of the accident. Here the officer is faced with the real need to understand the term "investigate" in order to trace out certain needed facts. To investigate is to search "minutely for the truth, also to inquire and examine with care and accuracy, or careful inquiry to find out what is unknown." This covers *what* the officer is supposed to do and *how* he or she is to set about the task of unravelling the problem at hand. Unfortunately, one cannot arrive at the accident scene and easily determine the guilty party. It requires time to make an investigation, so an officer has no reason for not making it a good one. There is a difference between "attending" and "investigating" an accident. Many reasons given for the "investigating" of an accident by an officer are merely the reasons for his or her attending, such as the preserving of peace and protection of life and property. Any officer "attending" an accident will direct traffic, care for the injured, and so on but make no effort to investigate.

The predominant cause of an accident is not always a violation. The officer should not narrow his or her work to a possible prosecution only but always check for *secondary causes* which might have been a contributing factor. What did the violation cause stem from—situation or circumstance? Ascertaining causes requires deliberate, painstaking effort; a lot of weighing, that which we sometimes call "expert opinion." The officer must use his or her expert opinion and intelligence. He or she may distinguish between provable opinion and provable fact, but the officer must be sure to include any opinions he or she has formed.

In view of the importance of accident investigation, it becomes apparent that the successful accident investigator must possess certain qualities and abilities. This will assure that the task is ac-

---

3. Ibid., pp. 27-28.

complished satisfactorily from the standpoint of the individual, the public, and the department which the investigator represents.

Characteristics of an Accident Investigator

The first of these is *enthusiasm*. The individual must be interested in his or her work, zealous and anxious to do the best possible job. Each new investigation must be accepted as a challenge. Lacking enthusiasm, the operation becomes routine, something to be accomplished with a minimum of time and effort. The completed investigation will fall short of the quality which is necessary. It will be a recording, not an investigation.

Determination is another essential for the successful investigator. He or she must possess tenacity of purpose to do a competent job, get the job done and done well or the investigation will lack that completeness which marks a good report.

The investigator must possess *sincerity*. He or she must truly believe in his or her work and its importance as a public service. His or her feelings must direct his or her actions toward a definite goal. Believing fully that what he or she is doing is worth doing well, he or she will be content with nothing less than sincere effort to achieve the best in each investigation.

A sense of *responsibility* is important—a desire to be an asset to the organization he or she represents. He or she must recognize the need to make a thorough investigation, do a proper job of enforcement, write a complete and concise report, and represent his or her department well at the accident scene. He or she should realize that investigation goes far beyond mere recording of readily available information on an official form.

The *ability to communicate* is essential. As an investigator, he or she will be expected to know more about the accident than a layperson. This information must be so written that any reader will have as much knowledge of the circumstances as the investigator. The investigator's report must contain all his or her knowledge of the accident. If, *after reading the report, the reader cannot ask a relevant question regarding the accident, the investigator has in some measure failed.* Either the investigation or the report has been in some way inadequate.

Proper representation of his or her department at the scene entails the ability to *represent authority* without a display of excessive command. The officer need not remind people who he or she is—they will recognize him or her the moment they see the uniform. The investigator is being observed all the time. Some welcome him or her, some believe in him or her, but others would like to criticize him or her. The actions of the officer reflect not only upon himself or herself, but also upon his or her department and the whole police profession.

The investigator must possess *investigative ability.* He or she must be able to interview, observe efficiently and completely, reach sound conclusions, and make his or her investigation meaningful and valuable. Some fatal accident investigations may be very difficult, as in those cases without survivors. With no witnesses, the accident must be reconstructed from physical evidence alone. There have been attempts to make it appear that victims of murder met death as the result of a traffic crash. Casual examination of the scene would have led to a false conclusion, allowing the murderer to escape. In many cases, however, proper application of investigative techniques have uncovered evidence that the incidents were not what they appeared to be and the responsible parties have been brought to justice. The ability to analyze situations—measure statements against physical evidence and probability—in order that his or her opinions will approach reality is essential for the investigator.

Dealing as he or she does with people who are not in a normal situation, the investigator must possess the qualities of *leadership.* It is necessary that he or she quickly bring order to the chaos existing at accident scenes. He or she must have the ability to control emotionally and physically upset people. He or she must use skill and tact in answering or asking questions. An attitude of disinterest or levity is out of place at the scene of any accident. No matter how commonplace or routine the particular incident may appear to the investigator, he or she must always bear in mind that to the participants there is nothing routine about it. To them it is an unusual, frightening, and sorrowful experience. The investigator must approach his or her task with empathy. As a further instance of leadership, the investigator must be able to recognize how much authority he or she can delegate to volunteers, as well as properly evaluate their actions to determine if they will actually assist rather than hinder him or her in the investigation.

Initiative is essential. Many times the investigator will not have a supervisor to direct him or her, so he or she must be able to think logically and clearly under conditions of duress and pressure. He or she must persevere to conduct a thorough and complete investigation with a minimum of guidance and direction.

A basic knowledge of *photography* will be of value. The investigator must plan his or her picture for its evidential rather than its news value. He or she should know where a photograph will help in his or her investigation and where one will not help.

A *sense of duty* to protect the rights of the participants is essential to the investigator. He or she must strive to get the facts. He or she must realize the need for conducting an impartial investigation without preconceived opinions to color or slant the investigation. It will help if he or she considers participants as things, materials, and formulas rather than people. He or she should approach the investigation in terms of what each vehicle did in relation to other physical conditions. With this viewpoint, he or she is less apt to be swayed by prejudices as to age, sex, race, color, occupation, or personality. Any of these may subconsciously affect the unbiased consideration of all the elements which he or she must collect and evaluate. His or her report of the investigation must reflect the true accident picture.

Knowledge of these as well as other police techniques and qualities will help. The successful investigator will always be alert for new knowledge, ideas, and techniques. When he or she uncovers a new way of doing things, he or she will pass the information on. Progress in any field is a group effort, built upon the contributions of many individuals. The accident investigator today profits from the experience of past investigators, and he or she, in turn, can contribute by making available to others the results of his or her experience and knowledge.

Procedural Steps in Accident Investigation

Of course one of the most important steps to developing a good accident investigator is the method in which he or she is trained. This may include one of several ways: training through a certified course at a local college or university; training taught to the individual at the basic or advance police academy, or by a police or traffic instructor during an in-service training session; and finally, training through actual field experience on a one-to-one basis.

However it is done, it must be done thoroughly and with great care in determining that all the important aspects have been covered. This should then be followed by field observation by the training officer or supervisor to validate the knowledge and skills learned by the officer. When this has been done to the satisfaction of his or her department, then the officer is ready to handle many of the routine types of accidents. The role of the individual from that of a basic police or traffic officer to that of an investigator is not an easy or fast transition. It will take many years of hard work, investigating many accidents in order to obtain that which we call "professional expertise." What should the investigator be familiar with in order to properly work an accident situation?

Every action of the investigator, from the original accident call to the completion of the report, falls within one of the three reasons for investigating accidents. Steps in normal accident investigation will be grouped under these three broad headings in the order in which they most logically fall.

In order to assure uniformity of understanding, it becomes necessary that certain terms must be narrowly defined. The following basic definitions are in common use in the accident investigation field.

There are three basic types of accident investigation. First, is the *"normal"* accident investigation which includes the investigation of those accidents where the principals are present, vehicles approximately as they came to rest, physical evidence unaltered, and with witnesses possibly present. Second is the *"late reported,"* or late investigated, accident in which knowledge of the accident is gained and the investigation, if any, is made after some lapse of time. Third is the *hit-and-run* accident in which the operator knowingly causes damage or injury and/or death and leaves the scene without identifying himself or herself.

A Fixed System of Operation

Some type of fixed procedure is extremely valuable in accident investigation, although it should not be a limitation of the investigator's actions. *It should be flexible.* Steps should be taken as conditions warrant, rather than by rote. By realizing the need for a fixed procedure, the investigator will use it as a sort of mental checklist. It will make the task easier for the officer—assist in determining the cause and gathering data, tend to make sure that nothing is overlooked in the investigation, give a definite starting point, and help to prevent duplication of effort. A planned procedure impresses the public with the efficiency of the investigator's department and builds favorable public opinion. Officers will find that accident investigation will consume a great part of their time. The better they know the correct procedure, the more reliable their reports will be, and the more pleasant their job will be. When one is familiar with a task, it is easier to perform.

Responsibility to Protect Life and Property

In a normal accident investigation, there are various steps which may be conveniently grouped under the broad heading of protection of life and property. Not all of these steps will be required in all accident investigations, nor will they always be done in any particular order. The important thing for the investigator to remember is that each investigation should be evaluated in the light of each of these steps, assigning top priority to that which in each instance is most important. The best procedure is a planned procedure which can be adapted to each event.

Before any investigation can commence, the investigator must be at the scene. Therefore, his or her *safe* arrival is essential. There are many reasons why the investigation should be made as soon as possible after the occurrence of the accident. It follows, then, that prompt arrival at the scene is also important. The first thing the investigator has to do is get to the scene *safely, yet quickly.*

Since the accident has already occurred, the investigator can do nothing to prevent it; he or she should take this fact into consideration and protect his or her life and others by driving at a safe speed. Taking unnecessary chances may lead to the investigator being involved in an accident himself or herself. This not only slows down the investigation of the accident he or she was assigned to cover, but also provides an additional accident which must be investigated. Bad public relations can be caused by the officer becoming involved in an accident while enroute to the original accident scene.

The arrival of the investigator is necessary to protect the scene and other traffic, preserve evidence which might be lost, and, in some cases, to save life by applying first aid. Another accident may occur because the highway is obstructed. Traffic congestion is usually present, and the investigator must attempt to restore normal vehicular movements. By prompt arrival, he or she will prevent driver irritation and poor public relations. To those involved in an accident, time is fleeting. Each minute they wait seems much longer than it really is. Also important is the fact that witnesses, who might be of great value, often become impatient and leave the scene once their curiosity is satisfied, and the investigator will not derive the benefit of information which he or she might otherwise receive.

A safe but quick arrival is a challenge to the investigator. One way is to *plan* the approach over the *quickest route* (not necessarily the shortest). The investigator must know the area—the shortcuts, detours, construction zones, and areas of congestion. A few seconds spent thinking out the best route may save minutes of hazardous high-speed driving. The officer should start at once upon receipt of the call. A short delay in getting started will have to be made up on the road, and a long delay means that time is lost which can never be regained.

Upon arrival at the scene, the investigator should *park safely and conveniently.* Consideration should be given to the safety of the officer, the patrol car, and other traffic. Parking opposite the accident, the traffic stream will have to be crossed each time it becomes necessary to return to the patrol vehicle. Parking on the same side of the highway will be safer and more convenient. By proper parking, the warning lights on the patrol car may be used to help protect the scene. Safety and availability of the equipment in the vehicle should be considered as use will be of the radio, flares, first aid equipment, and the like. Also, unguarded equipment may offer an incentive to potential looters. By safe parking, the investigator will help to establish a safe parking and passing pattern, and his or her task will be much easier if he or she has the various items needed close at hand and where they can be kept under observation.

It will be necessary to *protect the accident scene* to prevent the occurrence of further accidents and to preserve the evidence which the investigator will need. This may require placing flares at night or under conditions of poor visibility. Thought should be given to the proper location of flares to give approaching drivers a chance to slow down before running into the obstruction and to channel traffic into a safe path around the accident. It may be necessary to direct traffic around the scene. In many cases, bystanders will volunteer to assist. If their aid is enlisted they should be told what to do; otherwise, their undirected efforts may hamper rather than help the investigation. Attempts should be made to keep onlookers in a safe location. This at times poses an extremely difficult problem to the investigator.

Where gasoline or oil is spilled on the highway, it may be necessary to request fire equipment to wash it from the pavement, or highway crews may be needed to sand the area to eliminate the hazard. Where a fire hazard exists, as from gasoline, liquid petroleum gas, or similar fluids, care should be used in placing flares, onlookers should be warned of the danger of smoking, and, if necessary, the crowd should be cleared from the area.

Where there is injury, *first aid* will be necessary. In certain instances, this may be the most important thing for the moment and should be given without delay. Where injury is minor, protection of the scene may warrant top priority. Placing flares and directing traffic while a person bleeds to death would undoubtedly be the wrong order of operations, as would bandaging slight injuries before protecting the scene when a high-speed road was blocked in an area of low visibility. The investigator must consider each situation on its own merits. The most important thing should come first, regardless of its position on the checklist of procedures. An ambulance should be ordered, if required. This is one of the first considerations when there is serious injury. Where there is absolutely a fatality, the coroner must be notified. Any question to doubt as to whether the victim is dead or alive should be resolved in his or her favor. He or she should be treated as alive unless, by the very nature of the injuries, there can be no doubt that he or she is dead.

The investigator must *care for* the *property* of those injured. A tow car may be needed, and it should be ordered as soon as it can be conveniently done. If possible, the driver's preference as to towing service should be ascertained and the tow car of his or her choice ordered. When the driver is unable to care for his or her own vehicle and the investigator orders it stored for safe keeping, a report of stored vehicle must be filled out, and the contents of the car listed and inventoried, securing the signature of the tow truck driver on the inventory form as a receipt for the vehicle and its contents.

Evidence Collection and Finding the Cause

The remaining steps of the investigation come under the headings of gathering evidence for prosecution and determining the cause of the accident. These two objectives are common to accident investigations and cannot be separated. They go hand in hand. Evidence as to the cause will lead to the cause. From evidence and statements of witnesses and participants, the investigator must determine the cause of the accident. He or she must bear in mind that there probably will be more than one cause in any but the most simple type of accident; where more than one vehicle is involved, there may be one or more violations on the part of each driver.

There are *three factors* in any accident—the *highway*, the *vehicle* or vehicles, and the *people* involved. All three should be checked for evidence. Although it is impossible to get all the facts about everything, the investigator should strive to get all the pertinent facts. If a violation is indicated, an effort should be made to secure sufficient evidence to prove the elements of that violation.

Usually the officer will begin to form an opinion of what happened early in the investigation and will look for evidence to confirm it. At this point care must be taken not to overlook evidence which does not agree with the investigator's theory. Evidence which tends to disprove his or her first impression may be very important. Things are not always what they seem. The officer must never attempt to distort facts to fit his or her theory; the opinion must be adjusted to fit the facts. He or she must never be reluctant to accept negative or conflicting evidence, no matter how convinced he or she is that the accident happened in a different way.

Every fact, each statement, and every shred of evidence must be analyzed and evaluated in order to get a true picture of what happened. To start with, the investigator is faced with a result. Step by step he or she must work backwards from that result to find the causes. He or she must proceed from the "what" of a situation in order to determine, insofar as possible, the "how" and the "why."

A logical first step is to *locate* and *question* the *drivers* and *witnesses*. This is done by making a quick preliminary check to see that all drivers are present, then attempting to locate witnesses. The investigator should get the operator's license of each driver, determine which vehicle each was driving, then look for persons who might have seen the accident. Many persons may remain at the accident until the police arrive, then leave before the officer has had an opportunity to question them. An early search for these people may provide the investigator with information which he or she would not otherwise obtain.

The person who begins this search with the query, "Were there any witnesses to this accident?" will probably not be very successful in finding any. The word "witness" has an unpleasant connotation in the minds of many people. They have been conditioned by motion pictures and other types of entertainment to visualize a witness in court being badgered by the opposing attorneys to the point of acute discomfort, and they want no part of it. The indirect approach is inherently much more productive. The officer should find out from the drivers, if possible, who was the first one they saw or talked to after the accident, then approach these persons and ask them what they saw. They might say that the accident had already happened when they stopped, but they may be able to furnish information as to who was there when they arrived. Thus, by indirection, a witness may be discovered. Many times, while giving first aid or performing some of the other duties incidental to the investigation, the officer may overhear someone telling how the accident occurred. This person may or may not be a witness. He or she may have talked to someone who did see it, and in this way, too, a witness may be located. A check and recording of the license numbers of the vehicles parked closest to the scene may be valuable. Drivers of these vehicles are most apt to be the first arrivals. Even if witnesses cannot be found at the scene, contacting these drivers at a later date may develop the fact that one or more of them may have witnessed the accident. However, witnesses should be located as soon as possible. Not only are they prone to leave when the excitement is over, but also their statements may be valuable in offering leads to develop during the actual interrogation of the drivers. They may also suggest what physical evidence the investigator should look for.

In all contacts with witnesses or potential witnesses, the investigator must act with tact, diplomacy, and understanding. It must be remembered that he or she is asking for something, and that these people have no obligation to tell him or her anything. Their assistance may be very valuable, if freely given, but if coerced into making a statement, their feeling of resentment toward the interrogator will probably destroy the value of any statement he or she may extract.

Drivers must be thoroughly checked as to their *physical condition.* Intoxication is a state which often changes rapidly. Tests for intoxication should be conducted without unnecessary delay. The condition of the driver at the time of the accident is important, not the condition an hour or two after the accident occurred.

Early in the investigation, it should be *determined if* it is a *hit-and-run,* for the procedures in this type of investigation are different from normal accident investigation. Time is of the essence, and the sooner the hit-and-run procedures can be put into effect the more likely will be the chance of apprehension.

The *second factor* to be considered is the *vehicle.* Much evidence may be developed by thoroughly examining and testing the vehicles involved. It should be determined if all the visible damage was the result of the current accident. The exact location and extent of damage should be noted. Evidence of crumpled fenders and broken glass which does not tie in may be the result of a previous incident. Fresh damage, not connected, may lead to the solution of an earlier hit-and-run. Perhaps the driver's statement is not correct, and damage which he or she does not explain may lead to the true account of the accident. A check of the mechanical devices—brakes, lights, windshield wipers, turn indicators, and mirrors—should be made to see if they are adequate. In some cases, operator's licenses are restricted to the operation of vehicles equipped with certain safety devices. The officer should check to determine if safety devices are employed if required.

The condition of each tire on the vehicle should be noted and a check made of the front end, proving the damage is not such as to make the test impossible. The driver's statement may give the investigator a clue as to what to look for. Brakes may fail or be faulty or inadequate. Loss of control may suggest a worn steering assembly. A fresh break in the steering mechanism will probably be a result of the accident rather than a cause. Other points to check are lack of safety devices on commercial vehicles, inadequate lighting equipment, defective windshield and vision obscurements. Tire failure is often claimed as the cause of an accident, but many times it is a result rather than the cause. The investigator should know whether or not vehicle failure contributed. In the event of prosecution, the driver may claim vehicle failure as a defense. If the investigator has not checked, he or she will be unable to refute this type of argument. However, if a thorough check has been made, he or she will have the answers to any possible questions which may arise.

The third factor is the *surrounding physical conditions.* Any which might have contributed to the accident should be carefully observed by the officer: holes in the roadway surface; loose material on the pavement; the weather; visibility from the driver's path of view, including obstructions to vision such as hedges, trees, poles, signboards; traffic control devices; and warning signs. All observations should be made from the *driver's line of sight.* The operator of a truck ten feet above the highway has a far different field of view than the driver of a low-slung sports car. Skid marks, gouge marks, tire marks, and vehicle position should all be checked and the information recorded in notes at the scene, for both the accident report and possible use when testifying.

In addition to notes, the investigator should make a sketch to aid investigation and preserve information. This sketch should be drawn with north at the top and all wording oriented to this direction. The scale should be large enough to make it clear, yet small enough to show the whole area. Everything that might have a bearing on the accident should be included: length and position of skids; distance traveled after collision; width of streets, shoulder and roadways; distance and height of view obstructions; location of traffic control devices; distance from the pavement to any fixed object which was struck; victims or parts of vehicles; and debris. Distances should be measured, rather than estimated or stepped off.

The investigator should not neglect to locate and record the evidence that may be hidden beneath the vehicles. When making the sketch, vehicles should be treated as transparent, viewed from the top, and evidence should be recorded in its proper location.

Necessary photographs should be taken. They may be a great aid to the prosecution as they present physical circumstances of a case in a manner that is easily understood. Most people are visual minded. A photograph of large or perishable evidence is accepted in place of the original. Include such items as skid marks, footprints, or a body to show the location and type of injury. But before the shutter is snapped, the investigator should ask, "What will this tend to prove?" If the answer is "nothing," the photo should not be taken.

The photographer must be qualified. He or she must know what he or she is doing and why, for the more important the picture the stronger will be the attempt to discredit or disprove it. A commercial photographer is usually not satisfactory for several reasons. In addition to being expensive, he or she may not know what is required, is often reluctant to appear in court, and may not keep proper data to insure that the photograph will be admitted as evidence. If a commercial photographer is used, the officer should direct the photographer to take the type of picture the officer needs and record exposure and other related data. It should be remembered that these photographs are not for publicity but are evidence, as is the sketch of the scene. Valuable as they may be, the investigator should not depend on photographs only. They may be an important part of the case, but photographs alone do not constitute a complete investigation.

Pictures of many of the following subjects are often needed as evidence or to complete the records of a case: general scene from driver's viewpoint; point of impact; traffic control devices; skid marks showing length and direction; nature of roadway at location, showing defects, position of cars, victims, and parts of vehicles after impact, showing distances from point of collision; view obstructions or the lack of them; blood, flesh, hair, fabrics, scrape marks, and the like which are frequently useful in hit-and-run cases; tire prints; footprints; defects in vehicles involved, such as missing headlight, stickers obstructing driver's vision, trucks lacking turn indicators; the roadside, to show the kind of district; sagging springs of an overloaded vehicle; and the license number of the vehicle for identification.

The investigator should also make a sketch. Photographs do not show measurements or dimensions. The photo may be incorrectly exposed, blurred, or otherwise worthless. The accident sketch is the record of the investigator's intricate, personal examination of the scene.

An accident investigator should be aware of the following list of "don'ts":

1. Do not discuss the accident with anyone outside of your department or outside the line of duty.
2. Do not take sides. The investigator is an impartial fact finder.
3. Do not tell any participant he or she is either blameless or at fault. These are matters for adversary proceeding.
4. Do not tell anyone you will be a witness *for or against* him or her. You will be a witness for no *individual*. Testimony must be honest and impartial.
5. Do not mention garage charges, repair bills, or civil liability, as these are not part of the investigation.
6. Do not enter into any discussion of a probable or possible fine. This is a court function.
7. Do not be hasty in making decisions as to cause. Consider every possible cause, gather all the evidence, then analyze to exclude factors not involved.

With the information gathered, the investigator should now determine the cause or causes of the accident. He or she should be very careful in this, as it is one of the most important parts of the investigation. It should be approached in terms of the violation, and facts should be gathered to prove the essential elements of the action. It is quite possible that there may be one or more violations on the part of each driver.

What the investigator faces upon arrival at the scene is the result of a combination of circumstances. He or she must start from this result and work backwards as the investigation progresses, until at its conclusion he or she knows the cause or causes of the accident.

The Probable Result

Broadly speaking, accidents may be classified as either one of two types: collision or noncollision. The collision may be with another vehicle, train, bicycle, fixed object, pedestrian, street car, animal, or other type of object. Noncollision may be running off the roadway, or overturning in the roadway, or other noncollision, such as fire, falling from a vehicle, or injury within a vehicle. These are not causes of accidents, but the results of unexpected or unconventional behavior of one or more of the elements of the traffic at that particular time. The result may be called the *key event*, which is the most significant event that happens to each traffic unit involved, for example, collision with another motor vehicle, or running off the roadway.

Earlier we talked about causes and their relationship to various types of defects in preparation, driving strategy, or evasive action. The term "key event" or "the rule of first harmful event" has been applied to our list of growing terms. The "key event" has a direct relationship to the cause of a crash. We turn now to the various causitive factors in accident investigation.

The *direct cause* of the accident is the unconventional behavior of a traffic unit which leads directly to the key event. An example of this would be a vehicle that had skidded on a curve. *Mediate causes* are connected with the accident not directly, but through the direct cause. There may be more than one mediate cause. It is in the area of mediate causes that accident prevention can be effected; the part each factor in the accident plays in its occurrence, that is, the contribution of the vehicle, the driver or pedestrian, and the highway.

Just as the key event is the result of the direct cause, and the direct cause is itself a result of mediate causes, so are mediate causes the result of earlier and more remote causes. There may be many of these, and they may be very remote indeed, such as the habits and attitudes of the persons involved.

For example, in an uncomplicated accident a vehicle runs off the road and overturns, injuring the driver. In this case, the result is an injury accident. The key event would be skidding on a curve, which might have resulted from a combination of mediate causes, including a slippery highway surface, excessive speed, smooth tires, improper brake application, and an inadequate signing of the hazard. These in turn may have many *early causes;* rain, poor driver attitude, insufficient funds for adequate highway signing, construction and policing, lack of education of the driver as to proper care of equipment, and many more acts, neglects, or conditions which lead up to mediate causes. The area of improvement in all three phases of traffic supervision fall within the area of mediate causes. Many of the early causes also offer a field where improvement may be effected, but others are so remote as to be of little immediate concern to the investigator, although they may play a part in the total picture of traffic supervision.

Summary

We have noted many reasons for the necessity of a sound accident investigation program; however, we must recognize also that unless we prepare our police officers in the proper techniques of traffic accident investigation, no program, however elaborate, will succeed.

To do an effective job of traffic supervision, a complete analysis of the causes of accidents should be carefully studied. While in recent years the mileage death rate has shown a decrease proportionate to the number of motor vehicles and miles traveled, most states have been unable to reduce their overall accident problem. This could be due to a number of factors, but most predominately, the three "E's."

The accident investigator needs certain qualities in order to do an effective and efficient job. However, there is no substitute for carrying out fully all of the procedural steps necessary in a complete field investigation. First, it is necessary to recognize the three basic types of accident investigation: normal, late reported, and the hit-and-run. Although there should be a fixed system of operation in accident investigation, it must always remain flexible to accommodate any unusual problem that may arise.

It must be emphasized that in any motor vehicle accident there are three factors that, when all linked together, form the chain reaction leading to traffic mishaps. The three factors identified were the highway, the vehicle, and the people. Sufficient physical evidence must be gathered to identify each of these factors with some contributing cause that brought about property damage, injury, or death.

It is finally up to the investigating officer to zero in on the key event, analyze and/or reconstruct the accident, look at possible alternatives, make a decision based on the facts, and then submit a comprehensive factual report to his or her agency describing the resulting traffic accident.

DISCUSSION QUESTIONS

1. Describe in full the terms key event, direct, and mediate causes. Apply them to the principles of traffic accident investigation and analyze.

2. Discuss why it is so important to understand the various characteristics of accident causation. Identify some of the more prevalent types of accident causes.

True–False

1. Accident investigation is the cornerstone of the foundation of any program of traffic supervision.
2. During recent years, scientists have turned to the word "crash" in preference to the term "accident" in reference to traffic collisions.

3. Good traffic collision reporting is really of little value in the overall traffic program.
4. The *direct cause* of the accident is the unconventional behavior of a traffic unit which leads directly to the key event.
5. Measurements of fatal accidents require only approximate distances and, therefore, the pacing off of distances is acceptable.
6. An investigator's duty upon arrival is to protect the scene from further accidents before any other task is performed.

10: Special Accident Investigation Problems

The Hit-and-Run

With the hit-and-run accident, the officer is faced with a situation where he or she is really conducting two investigations. There is the investigation to determine the cause of the accident, and also the one to determine the identity of the driver who has left the scene. This is one type of accident where it is readily apparent that a violation has already occurred, that being evasion of responsibility. The officer, however, must not lose sight of the fact that leaving the scene is not a cause of the accident. The investigation, insofar as determining cause, must be complete and thorough in addition to locating the missing driver.

Hit-and-run may be a felony, as in cases resulting in injury or death, or may be a misdemeanor, wherein only property damage is involved. In either case, the accident should be investigated just as any other accident, except that the hit-and-run procedure should be followed.

The elements of felony hit-and-run, as established by the Vehicle Code, are as follows: a driven vehicle involved in an accident resulting in injury or death; failure to stop immediately (not instantly) at the scene and furnish identification by giving name and address of driver, name and address of registered owner of the vehicle, and upon request and if available, show operator's license; and failure to furnish aid, if requested or required.

To prove a case of felony hit-and-run, therefore, it is necessary to prove each of the foregoing elements. It must be shown that a driven vehicle was involved in an accident in which someone was injured or killed and that the driver had knowledge of the accident. Knowledge of the injury or death is not necessary. It must be proved that the driver *either* failed to stop or identify himself or herself, *or* offer aid if it was necessary or requested. The driver may stop or return to the scene, but unless he or she identifies himself or herself, the elements of the crime still exist. The investigator may prove knowledge, identity, and the offense by means of witnesses or entirely by circumstantial evidence. The owner of the vehicle, as a passenger, may be as guilty as the driver if he or she makes no attempt to stop at the scene and comply with the hit-and-run statute. The owner, under these circumstances, may be convicted of hit-and-run driving. Actual collision or contact between the vehicles involved is not necessary. "Involved in" is given the broader construction so that physical contact is not required. However, the driver involved must have committed the violation that caused the accident if there is no physical contact.

Time is important in hit-and-run investigations. The longer it takes to locate the driver the more time he or she will have to repair the damage, destroy evidence, and establish an alibi. Therefore, the investigator faced with a hit-and-run investigation should get information on the air for the benefit of radio-equipped units as quickly as possible so that a wide search for the violator may be instituted at once.

Procedure at the scene of a hit-and-run investigation should be keyed to this need. A check should be made, first for injuries, then to be sure that the drivers of all vehicles involved are present. As soon as it is determined it is a hit-and-run, a description of the responsible vehicle, the driver and occupants, direction of travel, area of possible damage, and probably paint transfers should be secured and the preliminary information broadcasted as soon as possible in order that a minimum amount of time is lost in launching the search.

Having arranged for this immediate lookout to be broadcasted, the investigator should question the witnesses fully. He or she should get their complete story and determine the amount of damage and injury. If there was a collision, how did the witness know? If a license number was obtained, how was it obtained and by whom? It is essential to maintain a chain of continuity of all evidence so the case will not break down in court.

The investigator should determine whether or not the suspect stopped at any time. If he or she did, did he or she converse with anybody? Any conversation should be recorded in the exact words, if possible, as an aid in questioning the suspect. Was any identification offered? Did the victim ask for any? The witness should be questioned as to whether he or she would be able to identify the driver. The best possible description of the driver should be obtained, as well as of the passengers. An accurate description of the vehicle should be requested, including the damage, make, model, year, color, and anything unusual or outstanding about the car. This should be as complete as the witness or witnesses can make it and supplemented by physical evidence the investigator discovers. Any new developments should be furnished to the radio dispatcher so that other units may be advised.

It should be determined whether or not the witness actually saw the accident. A person who did not see the accident itself might have seen the damaged vehicle leave the scene. Persons working in nearby areas or motorists arriving from the direction of the suspect's flight might come under this catagory. A thorough canvas of the area should be made for witnesses.

No detail should be overlooked. If there was an injury, was it apparent? Did the suspect make any attempt to assist the victim? Was the victim conscious? How long from the time of the accident was it before the investigator arrived? Were there other violations on the part of the suspect such as excessive speed or driving while intoxicated? This is a twofold investigation, and no particle of evidence should be overlooked.

The scene should be literally combed for physical evidence. Extremely small parts broken from the suspect's car may be matched with the vehicle to make a positive identification. Paint transfers or chips may be discovered. Clothing should be preserved to be analyzed, using great care not to handle clothing unnecessarily or roughly as paint is invisible to the naked eye in many cases and might be lost. The clothing of persons killed should be obtained from the coroner. In auto-pedestrian cases clothing should be analyzed whenever possible. Chain of possession should be maintained, and clothing should be handled as little as possible and properly marked for identification. Photographs of the accident scene, victim and vehicle should be taken, and other evidence photos obtained as indicated by the circumstances.

The investigator should be alert for the return of the responsible driver to the scene of the accident. Mingling with the crowd, even going so far as to follow the officer closely, the driver may attempt to ascertain the progress of the investigation. The inquisitive person who displays an unusual interest might well be the wanted driver trying to discover how close the investigation is coming to him or her, possibly going so far as to destroy evidence.

Follow-up activity will include checking garages, wrecking establishments, service stations, and repair shops and putting them on the alert for the wanted vehicle. Many other channels of group activity will be followed by other members of the department, as well as the investigator.

Should no apprehension of the suspect result, the investigator should revisit the scene periodically at the same time of day, day of the week, and date of the month. Through this pattern he or she may discover route salespeople, delivery drivers, or others who have a regular routine and may have been witnesses to the accident.

Once the vehicle is located and identification established, the driver will have to be found and connected to the crime. When located, the approach should be positive. The suspect may spontaneously make a statement which will identify him or her as the driver involved. However, officers must remember to properly admonish the suspect.

Hit-and-run investigation and apprehension is one of the most difficult and interesting jobs of the investigator. There is no apparent motive. The driver may have acted spontaneously through fear or nervousness to avoid apprehension because of being intoxicated, unlicensed, with "the other woman/man," uninsured, or wanted for another crime.

After apprehension is made, it is vitally important that the broadcast be cancelled and all alerted establishments notified. A word of thanks for their cooperation would not be amiss. They might prove useful in a later case.

Good hit-and-run investigation and reporting is merely digging a little deeper to obtain more facts. It is the small, easily overlooked details which, if found and checked, would start the chain of events leading to apprehension of the violator. Success in this field is a group effort, calling for much cooperation with other agencies as well as within the department. Cases are solved by common sense, hard work, and a small amount of technical knowledge.

Establishing Traffic Accident Violations

One of the primary purposes of a traffic accident investigation is to obtain evidence of violations that have occurred and record the information in such a manner that a complaint can be issued against the violator. Uniform Vehicle Code sections are shown below in modified form to assist officers in recognizing the necessary elements. A complaint cannot be obtained if one or more of the elements is missing or cannot be proven. It is not necessary, however, that any one person witness *all* of the elements. A conviction can be sustained, for instance, where several witnesses contribute only portions of the elements individually, but together can supply all of the elements. To illustrate, this procedure is often used in drunk driving cases, with police officers testifying to intoxication only, and other parties or witnesses testifying to driving only.

When interviewing parties and witnesses at the scene of a traffic accident involving any traffic violation, the investigating officer *must* question each party or witness concerning *each element* of any suspected traffic violation. How each element of the intended prosecution can be proven should be covered in the opinions and conclusions section of the particular state's Accident Investigation Form. *No element* should remain *unresolved;* every person interviewed should state:

1. That the element *was* present.
2. That the element was *not* present.
3. That he or she does not know whether it was present or not.

The "Investigation" portion of the following material does not include the basic information needed for every accident investigation, such as direction of each vehicle, speed of vehicle, location when danger was first noticed, location of witnesses, or other general information, unless it is specifically needed as an element to obtain a conviction of a particular vehicle code section.

Hit-and-Run Felony

The driver of any vehicle involved in an accident resulting in injury to any person other than himself or herself, or death to any person shall:

1. Stop immediately at the scene.
2. Comply with his or her state's Vehicle Code requirement dealing with injured persons resulting from a traffic accident, i.e.,
   a. give name and address,
   b. registration number of the car,
   c. name of the owner of the car,
   d. exhibit his or her operator's license, if available, to the person struck or any police officer, if requested.
3. Render reasonable aid to the person struck.

NOTE: Failure to comply with *any* of these elements would constitute a violation of this section.

Investigation

In addition to the above elements (although not an element by statute), it is difficult to obtain a conviction without proof of guilty knowledge of the accident. Therefore, an investigator should gather evidence to show that:

A. The suspect knew or should have known of the accident because of extensive damage, loud noise, or other unusual circumstances.
B. Did the suspect stop? If so, where? Did he or she leave and then return? Why did he or she leave? How long was he or she gone?
C. If he or she stopped, did he or she give:
   1. Correct name?
   2. Correct address?
   3. Registration number of his or her vehicle?
   4. Name of the registered owner?
   5. Help to the victim?
   6. Did he or she exhibit his or her operator's license? (Was it requested?)
D. What did the suspect do that would indicate that he or she knew that he or she had been involved in an accident? Did he or she:
   1. Slow down and look back?
   2. Stop and look back? Walk back, look, then leave?
   3. Leave, then drive past again without stopping?
   4. Swerve, or take evasive action?

Felony Drunk Driving

To sustain a conviction of this section, the following four basic elements must be present, that is, an investigation must cover each in order to be complete.

Any person who:

1. While under the influence of intoxicating liquor or under the combined influence of intoxicating liquor and any drug,
2. drives a vehicle,
3. and does any act forbidden by law, or neglects any duty imposed by law,
4. which act or neglect proximately causes bodily injury to any person other than himself—is guilty of a felony.

NOTE: Driving a *motor* vehicle is not a requirement. The act or neglect must "immediately precede and produce" the injury. All reports should show a strong relationship between the proximate cause and the injury.

Investigation

A. Suspect
   1. What have you been drinking? How much? Where? When?
   2. Where were you going when the accident happened? (Establish driving)
   3. How did the accident happen? (Establish violation)
   4. What injuries were sustained? By whom?

B. Other parties and witnesses
   1. Has he or she (suspect) been drinking? Why do you think so? What is your opinion of his or her degree of intoxication? Has it affected his or her ability to drive?
   2. Was he or she driving the car? Did you see him or her in the driver's seat shortly after the accident?
   3. Did he or she commit any traffic violations? Were they the cause of any injuries?
   4. Who was injured besides the driver?

Misdemeanor Drunk Driving

It is unlawful for any person who is:

1. under the influence of intoxicating liquor,
2. or, under the combined influence of intoxicating liquor or *any* drug,
3. to drive a vehicle on a highway.

This charge is valid when:

1. The arresting officer witnesses the arrestee commit the elements of driving on a highway while under the influence of intoxicating liquor *or* combined influences, or
2. A witness can establish the element of driving on a highway, and the arrestee is intoxicated at the time of arrest.
3. The element of driving can be established by circumstantial evidence.
4. Arrest without warrant by a peace officer out of an accident.

NOTE: There are numerous degrees of intoxication from "barely under the influence" to "falling down drunk." The word "intoxicated," rather than the term "under the influence of intoxicating liquor," *should* be used in reporting officers' opinions.

Investigation

A. Parties and witnesses
   1. What have you had to drink? How much? When? Where? If you have not had anything to drink, why can't you do

the balance tests?

2. Where were you driving? How did the accident happen?

Narcotic Driving

It is a felony to drive a vehicle under the influence of narcotics. It is unlawful for any person who is:

1. addicted to the use, or
2. under the influence of narcotic drugs
3. to drive a vehicle upon a highway.

NOTE: The suspect should be taken to a specialist or doctor to obtain expert opinion of the cause for objective symptoms. No accident or injury is necessary to investigate. The charge is not valid if the driving is on private property, but the Health and Safety Code is still applicable.

Investigation

A. Suspect
   1. Why are you unable to perform balancing test? What medicine are you taking? What is your doctor's name? How long since your last "fix?" Check for needle marks, "hype kit," and so on.
   2. Where were you driving the car? From where?

B. Parties and witnesses
   1. Are his or her actions normal? Does he or she appear to be under the influence of anything? If so, did you notice the odor of alcohol? (To disprove drunkenness)
   2. Did you see him or her driving? Where? Was he or she in the driver's seat?

Nonnarcotic Drugs

Driving a vehicle under the influence of nonnarcotic drugs may be either a felony or a misdemeanor depending on the individual state laws. Any person who:

1. while knowingly under the influence of any drug other than a narcotic;
2. to a degree which renders him or her incapable of safely driving a vehicle;
3. drives a vehicle;
4. does an act forbidden by law or neglects any duty required by law;
5. which proximately causes injury to another person.

NOTE: It is not necessary to be a narcotic, not required to be a *motor* vehicle, and not necessary to be on a highway. The act or neglect must "immediately precede and produce" the injury, and the injury must be to someone other than the driver. Death is not included in this section (use manslaughter charge). Clearly establish the traffic violation; the charge may later be reduced.

Investigation

A. Parties and witnesses
   1. Does he or she appear normal? Do you think he or she is under the influence of anything? Why?
   2. Is he or she a safe driver? Would you ride in a car with him or her driving?
   3. Did you see him or her driving? Was he or she in the driver's seat?
   4. Did he or she do anything illegal that caused the accident? Any traffic violations?
   5. Was anyone besides the suspect injured?

B. Suspect
   1. Are you under a doctor's care? Doctor's name? If so, what drugs are you taking? Where are they? How many? How long?
   2. How did you feel just before the accident? Have you felt that way before? (Safe driving?)
   3. Where were you driving?
   4. How did the accident happen?
   5. Who was injured? What were the injuries?

Left Turn at Intersection

Requires a left-turning vehicle to yield the right-of-way to vehicles that:

1. have approached an intersection (waiting at red signal), public or private property, or an alley, or,
2. are approaching and are close enough to constitute a hazard at any time during the turning movement.

NOTE: Right-of-way violations occur when there is "danger of conflict" because this section involves a left turn. Other Vehicle Code sections may be involved, such as wrong lane, required signal from 100 feet, driving left of double lines, and nonturner yield.

Basic Investigation

A. Approach to intersection, public property, or an alley
   1. Lane traveled
   2. Speed (mph)

B. Signaling
   1. Type of signal, hand or mechanical
   2. Distance prior to turn, signal was given

C. Arrival at turning point
   1. Determine if party stopped or rolled into turn (mph)
   2. Locate point in relation to where party started his or her turn

D. Yielding (or lack of same)
   1. Describe number, speed, and location of cars yielded to
   2. Describe speed and location, approaching vehicle (party)

E. Path of travel following turning point
   1. Note position of point of impact, skid marks, and obtain statements relative to same.

Left Turner

A. What street were you on? Which direction were you going? Which lane? What speed? (Set the picture)

B. Did you signal? How? Where were you at the time of start of signaling? (Turn signal violation)

C. Where was the other car at the time you started to turn? How fast was it going? How fast were you going then? Did you slow or stop? Where? Where did your turn actually start? Did left turner stop, then restart?

D. Did any other cars pass going in the same direction as the other car? How many? How fast? Were you partially in a turn while you were waiting? How long did you wait there? Where was the other car when you first arrived in turn position?

Nonturner

A. Which direction were you traveling? How fast? Which lane?

B. Where was the other car when you first saw it? Its speed? At the time it *first* started to turn, where were you? Your speed? If violator stopped, then restarted?

C. How long was he or she stopped? Where were you at the time you first observed other car stopped? Were there any vehicles in front of you? How many? How far ahead?

Red Signal

Traffic facing the red signal shall:

1. stop at a limit line if marked,
2. otherwise before entering the crosswalk,
3. or if none, before entering the intersection.

This section requires drivers to stop at certain locations when a red signal is displayed. Therefore, parties and witnesses should be interviewed to find where the violator's vehicle was at the time the red light came on and whether he or she stopped for it.

Investigation

A. Both parties
   All parties should be asked the following questions in relation to every party involved:
   1. Which direction were you traveling? How fast?
   2. Where on the roadway were you when first observing signal? Which signal was observed? What color was it?
   3. What color was the signal when the party entered the intersection?
   4. Did the signal change? To what color? Where were the parties then?
   5. Where was the other party when first observed? What was the condition of the signal then?

B. Witness
   1. Where were you at the time of the accident?
   2. What attracted your attention to the accident?
   3. Which party entered the intersection against the red signal?
   4. Which signal did you observe? What color was it?
   5. How fast were the parties traveling?
   6. Do you know either party?

If witness' attention was attracted to accident by the sound of the impact:

1. Did you look at the signal? What was the condition of the signal? Which signal was observed?
2. How much time elapsed between the time of the accident and the time you first observed the signal?
3. Did the signal change after the accident? How long after?
4. What was the location of the offending vehicle when first observed and the condition of the signal at that time?

Following Too Closely

The driver of a motor vehicle shall not follow another vehicle more closely than is *reasonable* and *prudent*, having due regard for the

1. speed of such vehicle,
2. traffic,
3. and condition of the roadway.

NOTE: Primary considerations must show that both vehicles were *in motion* at the time of the collision or that the front vehicle had been *stopped for not more than one second.* This is essentially a speed and distance violation, and the investigation must show a hazardous condition—following too closely for a period of time.

Investigation

A. Rear car driver
 1. What was the speed of the front car? Your speed?
 2. How far behind were you? (Feet or car lengths)
 3. How far did you follow at that spacing?
 4. Condition of the road surface?
 5. Did the other person signal?

B. Front car driver
 1. What was your speed before the accident? What was his or her speed? (If observed)
 2. What atrracted your attention to the vehicle behind you?
 3. How far behind was it? How far did he or she follow there?
 4. Condition of the road surface?
 5. Why did you slow or stop?
 6. What signal did you give?
 7. How long were you stopped before being struck?

Yield Right-of-Way After Stop

The driver of any vehicle shall *stop* at the entrance to a through highway and shall *yield* the right-of-way to *other vehicles* which have:

1. entered the intersection—or
2. are approaching so closely as to constitute an immediate hazard.

NOTE:

A *stop* is required before this section becomes operative. If the driver fails to *stop*, use stop sign section. "Immediate hazard" is determined by establishing the distance and speed of the approaching vehicle at the time the starting vehicle is just starting into the through highway.

Investigation

A. Violator
   1. Did you stop? Where? How long were you stopped? (Seconds) Did any vehicles pass you on the through highway?
   2. Where was the other car at the time you were *just stopped*? Its speed?
   3. Where was the other car at the time you were *just* starting into the intersection? Its speed? Your speed?
   4. (If skids are observed leading into the intersection, are those your skids?)

B. Other party
   1. Where was the other party when first observed?
   2. Did he or she *stop* before entering the intersection? How *long* was he or she stopped? If not, his or her speed?
   3. Where were you at the time he or she was just starting into the intersection? Your speed? His or her speed?

Unsafe Turn or Movement

No person shall:

1. turn a vehicle from a direct course,
2. or move right or left upon a roadway until it can be done safely, and
3. then only with an appropriate signal if another vehicle is affected.
4. The person must signal during the last 100 feet traveled.

NOTE: Even if a signal is given properly a turn is still prohibited until it is safe to do so. Do not use this section for unsafe lane change on laned streets; it generally applies to lane changes within intersections.

Investigation

A. Violator
   1. Which direction were you going? Speed? What portion of the roadway were you traveling in?
   2. Did you signal? How? Window up or down?
   3. Where was the other car when you first saw it? Speed?
   4. Where was it when you started to turn?

B. Other party
   1. Which direction were you going? Speed?
   2. Where was the other car when you first saw it? Its speed?

3. Did he or she signal to turn? How?
4. Where were you when he or she started to turn? Your speed at the time?

Unsafe Start or Backing

It is unlawful to:

1. start a vehicle
   a. stopped,
   b. standing,
   c. or parked,
2. or back a vehicle
3. until such movement can be made with reasonable safety.

NOTE: This section is usually used for drivers suddenly pulling out from the curb or backing unsafely on a street—do not use it for backing out from a private driveway (some state vehicle code sections apply to vehicles entering a highway from private property whether entering forwards or backwards). Also note that signaling is not specifically required.

Investigation

A. Violator
   1. Where was the other party at the time you first started to move? Speed?
   2. Did you have sufficient time to complete your movement safely?
B. Other party
   1. Where were you when his or her vehicle first started to move?
   2. How fast were you going at that time? (Pedestrians—walking or running?)

Unsafe Lane Change

Whenever any roadway has been divided into two or more clearly marked lanes in one direction, the following rules apply. A vehicle shall be driven as nearly as practical

1. entirely within a single lane,
2. and shall not be moved from such lane until the movement can be made with safety.

NOTE: That street must have at least two lanes in one direction and must be marked. Also the vehicle need *not* actually *change* lanes—merely "move from" is sufficient (this includes "starting" to change lanes then moving back).

Investigation

A. Violator
   1. Which lane were you in?
   2. Did you signal prior to attempting to move from the lane?
   3. Were there any vehicles in the lanes on either side?
   4. Where was the other vehicle at the time you started to leave your lane?
   5. Did the other vehicle change direction or speed when you started to leave your lane? Why?

B. Victim
   1. Where was your vehicle when the other vehicle first started to turn? His or her speed? Your speed?
   2. Did he or she signal prior to the turn?
   3. Did you swerve or slow to avoid him or her? Why?

Pedestrian

Every pedestrian crossing a roadway at any point other than within a marked crosswalk or within an unmarked crosswalk at an intersection shall yield the right-of-way to all vehicles upon the roadway.

NOTE: Traffic signals are not required at adjacent intersections.

Investigation

This section essentially is a speed and distance problem. Parties should be interviewed to determine:

1. That the pedestrian was in the roadway between intersections but not in a crosswalk.
2. That a hazard was present *due to the relative positions* of the vehicle and the pedestrian.
3. That the pedestrian *did not yield* resulting in death, injury, or property damage.

Questions

A. Violator
   1. Where are you going? From where?
   2. Where was the vehicle when you stepped onto the roadway? Speed of the vehicle?
   3. Did you yield right-of-way to any vehicle?

B. Driver of vehicle
   1. Where were you when the pedestrian entered the roadway? Your speed?
   2. Was the pedestrian running or walking? Did he or she stop for you?

3. Was your attention distracted? Could you have stopped prior to the traffic accident?

Investigation of Brakes as Unsafe Equipment

This information is presented to acquaint the officers with the procedures necessary to initiate filings for violations of the Vehicle Code section dealing with brakes. The procedures to follow regarding brake investigations of vehicles with airbrake systems is also included.

Presentation

I. Procedure
   A. Thoroughly inspect the braking system and endeavor to establish whether or not the brakes were faulty *prior* to the traffic accident. The following procedure has been prepared for use in checking braking systems for defects.
      1. Determine if the system is hydraulic or mechanical.
         a. Whether hydraulic or mechanical, attempt to depress the foot pedal to the floorboard by fingertip.
            1. If this can be accomplished and the system is hydraulic, determine if the pedal can be restored to normal by pumping it. If so, record the number of strokes required.
            2. If the foot pedal can be depressed to the floorboard by fingertip and the system is mechanical, inspect the linkage for defects or improper adjustments.
      2. When inspecting hydraulic systems, check the area in the vicinity of the master cylinder (usually on the engine side of the firewall on the later models) for evidence of brake fluid leakage.
         a. Check for leaks in the hydraulic lines.
         b. Check the inside portions of the wheels where evidence of leaking wheel cylinders would be present.
         c. Check for evidence of leakage on the roadway.
            1. Inspect that portion of the roadway over which the vehicle passed just prior to the place of impact.
      3. Take photographs that will illustrate or substantiate

any faulty or unsafe condition of the braking system.

4. When in doubt if the braking system is unsafe, conduct a stopping-distance test, if feasible.
    a. The test speed shall be 20 mph and the legal stopping distance shall not exceed 25 feet.

NOTE: The preceding values apply only to passenger motor vehicles that are designed to carry not more than nine persons including the driver. (Refer to the respective state code for the legal stopping distances for motor vehicles).

5. Ascertain if the involved vehicle is equipped with an adequate emergency or parking brake.
    a. It must be capable of locking the rear driving wheels and holding the vehicle stationary to the limit of traction of the brakes wheels.

II. Reporting Requirements

A. The success of the prosecution rests upon the ability of the investigator to establish the essential elements of the offense and to make a comprehensive report of all facts revealed by the investigation. It is mandatory to establish two elements in order to request a filing of the brake section.

1. Establish that the braking system of the involved vehicle was *faulty* or in an *unsafe condition.*
    a. The accident investigation form should include a complete, comprehensive statement of facts revealed by the investigation of the braking system, including the results of any tests that were conducted.
2. Establish that the *driver* of said vehicle had *knowledge* of the unsafe condition of the braking system *prior* to the traffic accident.
    a. The driver's statement on the accident investigation form shall contain positively or negatively his or her knowledge of the unsafe condition of the brakes prior to the accident.

Witness and Driver Statements

Taking statements from drivers and witnesses is an important part of many investigations. This may simply consist of asking what happened and recording it in the investigator's own words or an informal, written statement signed by a witness. In serious cases, it may be a formally recorded statement taken by the prosecutor and stenographer. Statements written and signed by drivers in required reports are confidential; therefore, drivers' statements should not be written or signed by them.

Statements are of value for several reasons. They give the investigator a clue to what happened and in which direction he or she should start the investigation; they tend to dissuade a person from changing his or her story at a later date and may, in some cases, be used to impeach a witness. Even if the person has not signed the statement, the fact that he or she has given it to the officer may keep him or her from changing his or her story. The statement may also be used to refresh the memory of a witness.

The investigator will take statements to assist him or her in determining facts. By statements, he or she may place the driver behind the wheel, determine probable point of impact if there were no skid marks, or tie skid marks to a particular vehicle, and establish other circumstances of the accident.

One of the first considerations in taking statements is to talk to the person in private. Many people are reluctant to talk if others are listening. They may be afraid of what the driver or his or her friends might do if the statement is unfavorable. Spectators may interrupt by interjecting their own ideas, and arguments may start as the result of disagreements between participants.

If possible, the investigator should have the witness or driver tell the story from his or her general location at the time of the accident. This will serve as a check on his or her reliability and establish whether or not he or she could see what he or she claims. He or she should be allowed to tell his or her story without interruption, except as necessary to keep him or her from rambling too far from the subject. The investigator must be sure he or she understands what the person means. For example, "I was passing several cars" could indicate the speaker was either meeting or overtaking other vehicles. Words do not always mean the same thing to the speaker and listener.

Questions should be objective, positive, and specific; not "Where do you live?" but "What is your address, please?" This will save time and make statements concise and pertinent. Statements should be verified with facts and discrepancies pointed out. No attempt should be made to "coach" the witness, but by indicating points of disagreement the person may adjust his or her recollections.

After the person has finished his or her statement, if there are any points he or she has not covered, specific questions should be asked. When the investigator is sure the statement will be complete, he or she should write it out and have the witness sign it. In some cases, it may be appropriate to have a witness write his or her own statement to keep him or her occupied and from discussing the event with others. Generally, however, it is better for the investigator to write the statement and have the subject sign. Some people have difficulty putting their thoughts on paper, and their statements will be inadequate. By writing the statements himself or herself, the investigator can be certain it is complete.

When writing the statement as the subject tells it, the investigator may consider deliberately including a few errors in minor details. The subject should correct these when he or she reads the statement and initial the correction. If he or she fails to notice the errors, they should be called to his or her attention. He or she should then correct and initial them. This precludes the chance of the subject saying he or she did not read the statement, and by having the pencil in his or her hand and becoming used to it he or she is conditioned toward signing the statment when requested.

If the witness refuses to sign, the officer must not insist upon a signature. He or she should use a positive approach in attempting to get the signature and, if this fails, drop the matter. The signature in itself is not essential. Having given the statement to an officer, the subject will be less likely to change his or her story if he or she knows it has been recorded.

While listening to a witness' statement, the investigator should not be in a hurry to start writing. Some people are reluctant to talk if they know the conversation is being recorded and will not give a statement. Also, the first time they make it, the statement may not be adequate and will have to be rewritten, which is time consuming.

Ordinarily, by the time the investigator is ready to take the statement, he or she will have a good idea as to cause and violation. He or she should be sure the statement covers all the necessary elements. Care should be taken to cover who was driving, direction of travel, approximate speed, condition of pavement and weather, and to tie in the skid marks, if any, with the appropriate vehicle.

A statement which does not add to the investigator's knowledge of the accident is wasted time, whereas a properly taken statement may prove of great value to the case and should be preserved. Witnesses have been known to change their stories. If a person who has given a statement in which he or she declares he or she knew nothing about the accident should turn up as a witness, the statement taken by the officer would do much to discredit the later testimony.

Questioning Drivers and Witnesses

To assist in routine investigations, the following questions should be asked in most cases. The answers to these questions should give the investigator a fairly good picture of how the accident happened.

1. *Exactly where were you when you first saw the other car?* The object of this question is to establish the first point and time of awareness. Therefore, the officer may have to say *heard* or even *felt* instead of *saw*. (The first point and time of awareness may be a jolt created by hitting a hole in the road or a soft shoulder). It is better to get the driver to show the location than to tell it. It is a good idea to establish the spot so the officer can locate it later by measurements. From this spot, the officer should look for himself or herself to visualize what the driver saw.

2. *What were you doing at the moment you first saw the other car?*
   First, let the driver tell it in his or her own words. If he or she is not clear, then ask him or her about two things:
   a. *His or her movement*—speed, direction, slowing down or speeding up.
   b. *His or her attention*—where had he or she been looking? Had he or she been talking, changing the radio, or something else?

3. *Where were you when you first realized you were in trouble?*
   This question is to establish the *point of surprise*. Again, have him or her show the location if possible.

4. *What were you doing at the moment you first realized you were in trouble?*
   This question also involves establishment of movement and attention. Do not ask it if question 2 was answered satisfactorily.

5. *Exactly what did you do, if anything, to avoid the accident?* Here the officer tries to establish possible evasive action. If the response is slow, do not urge or coach an answer. If coached, the officer may get what the person thinks he or she *should* have done and not what he or she *actually* did. This is why "if anything" is included in the question.

6. *Then, what happened?*
   Let the driver describe the actual event in his or her own words.

7. *Exactly where did the collision take place?*
   This question need not be asked if the *point of impact* is readily apparent due to physical evidence.

8. *Where did you stop after the collision?*
   Again, not necessary if indicated by physical evidence. However, vehicles are often moved prior to the investigator's arrival and establishment of *point of rest* is often necessary to a valid investigation.

9. *What is the first thing you remember after the accident?*
   Take the time to allow an answer in his or her own words. Do not press for an answer. If the driver was knocked unconscious, this question is vital. The next logical question would be:

10. *What is the last thing you remember clearly before the accident?*
    Do not appear surprised at anything he or she says. However, if unconsciousness was a result of the accident, attempt to verify his or her statement by a statement from an unbiased witness.

11. *Whom did you first see after the accident?*
    This may be a good clue to an eyewitness. Witnesses have a habit of "not wanting to get involved."

12. *What did you say to him or her?*
    Here the object of the question is to get an idea of explanation of the accident the driver gave immediately following the crash. Encourage him or her to say more by asking, "What else, if anything, do you remember?"

13. *What did he or she say to you?*
    This is a means of checking to see if the drivers have changed their stories by comparing this with what the other driver has already told the officer.

As in any criminal investigation, if possible, the parties involved should be separated when these questions are asked. If an officer attempts to question them together, he or she stands a very good chance of ending up in the center of an argument.

When asking these questions, the officer should be sincere. Although it may be another routine investigation to him or her, this may be the most important part to the people involved. Most people are not involved in many accidents and a $25 dent in their fender can be very important to them. If the officer handles each accident investigation as a professional, he or she will find that the answers he or she seeks will generally appear. These questions are by no means all that may throw light on what happened. They can be adapted to meet the needs of any particular incident. They do, however, answer the important questions of who, what, where, why, when, and how.

Enforcement Action

Having gathered the evidence which will support a prosecution, appropriate enforcement action is the next step. In most traffic accident investigations, the offense involved will be an infraction in which case the officer will be precluded from taking enforcement action at the scene due to the legal restraints imposed upon him or her. In cases where the evidence is clear cut and apparently sufficient to support a charge, the investigator should inform the violator of his or her intention to file a complaint. In cases where the prosecutor desires the opportunity to review the evidence before issuing a complaint, the officer should take a positive stand. If he or she believes there is sufficient evidence, he or she should tell the violator that it is his or her intention to ask for a complaint. This paves the way for psychological acceptance on the part of the violator of subsequent notice that a charge has been filed. If the officer attempts to evade taking a stand by advising the driver that he or she will make further investigation, confer with his or her superiors, and so on, and a complaint may be filed, the door is opened in the violator's mind for a not guilty plea. He or she may feel that the officer is not certain in his or her conclusions and may be reluctant to accept his or her guilt to the extent he or she would if the officer took a positive stand. This points up the fact that an investigation must be complete, thorough, and conclusive.

Should the prosecutor wish to review the evidence before approving a complaint, the violator should be notified as soon as the complaint is filed. If the service of a warrant is his or her first notice of the court action, good public relations suffer. When the officer is empowered by law to take enforcement action at the scene, an arrest is in order.

It is good practice for the investigator to review the entire investigation just before leaving the scene. This serves two purposes. It gives him or her a double check to assure that all points have been covered and that all of the necessary information has been gathered. A physical re-check may disclose a piece of overlooked evidence, and property of the participants may be found which might otherwise be lost. If two officers are conducting the investigation, they should compare notes to be sure that each has all the necessary information. Reviewing the investigation takes a little time but may be very productive in terms of results obtained.

One possibility the investigator should not overlook is that of the "staged" accident designed to defraud insurance companies by means of nonexistent injuries supposedly sustained in a carefully engineered "traffic accident." Some individuals, including attorneys, doctors, and garage proprietors, have conspired to collect damages for fictitious injuries by carefully staging traffic accidents. This type of fraud is difficult to detect. Officers should carefully observe the demeanor of persons involved, developing good factual detail and watching for inconsistencies or fabrications. If fraud is suspected, careful questioning should be carried on to determine its motive. It may vary from the desire for easy money to outright murder.

Some of the circumstances which might suggest possibility of fraud are lack of witnesses, no plausible reason for the accident, one person freely admitting liability while the other seems to be feigning injury. "Victims" may prefer to seek their own medical care in preference to usual facilities. While these are not conclusive evidence of fraudulent attempts, officers should not ignore the possibility that such a conspiracy might exist.

Before leaving the accident scene, the investigator must be sure that it is in safe condition: vehicles and debris must be cleared from the highway, spilled oil sanded or covered with dirt, flares extinguished, and as much evidence of the accident as possible removed. Otherwise, traffic may become involved in additional accidents due to sudden slowing or stopping as drivers observe signs of an accident.

Often it will be impossible to complete the investigation at the scene. It may be necessary to contact additional witness to determine the true causes. Driver or other persons may have been injured and taken to hospitals or their homes after treatment. Sometimes the officer will do this further investigation himself or herself, and in other cases it will be done by officers assigned to accident follow-up. Should it become necessary to interview injured persons, the officer should check with the doctor in attendance before attempting to talk to these people. In some cases their condition will be such that it will not be advisable to interview them. In other cases, they may have been given sedatives or other treatment so that their statements will not be reliable. Each instance will have to be treated on its merits by the investigator. Just as no two accidents will be alike, so no two investigations will be alike. The planned procedure must be flexible, and steps taken in the order called for by each set of circumstances encountered.

The final step of the investigation will be that of making out the accident report on a standard form. In filling out the report, the investigator will be guided by his or her departmental policy, bearing in mind that to be a good report, it should leave no logical questions unanswered which might be asked by any of the several agencies which will use it as a source of information. A good investigation will lose much of its value if the information is not transmitted by means of a thorough, accurate, complete report, including in total the who, what, when, where, why, and how of the accident. Similarly, without a thorough, accurate, and complete investigation there can be no good report.

In addition to normal accident investigations and the hit-and-run investigation, the third type will be encountered—late reported accidents. Late reported accidents are those which come to the attention of the investigator after some lapse of time. Usually the vehicles involved are no longer at the scene, and much of the evidence will probably have been altered or destroyed. The degree of follow-up of these accidents will usually depend upon the severity. Fatalities will require a complete investigation. A considerable amount of investigation may be done where serious injury is involved, whereas minor property damage accidents may only be recorded on the official forms as simply a driver's report.

Summary

In this chapter, we have focused on some very critical issues such as the hit-and-run violator, methods and evidence needed to establish uniform vehicle code violations, taking of drivers' and witnesses statements, and finally the enforcement applied to accident investigation.

Space does not permit us to go into greater detail on the investigation of hit-and-run accidents. What has been discussed, however, are the essential methods for the field investigator to apply during the course of such an investigation. The investigator should understand that the hit-and-run suspect is not someone who merely left the scene by some unfortunate act which he or she was unable to control, but rather by impulse and by design. The field investigator must exercise every immaginative tool at his or her disposal in an attempt to piece the hit-and-run puzzle together. This task is not easy and sometimes may take days, weeks, or months to solve. By using some of the available physical evidence at the scene, statements of other drivers and possibly some disinterested witnesses, the investigator must work diligently in reconstructing the events that took place. Every possible detail must be examined and none should be overlooked.

There has always been a definite need for some uniform guidelines to establish traffic accident violations, and the material in this chapter should be helpful in providing some of these. It is imperative that the police officer, state trooper, county sheriff, highway patrol officer, or state police officer fully investigate each traffic crash with the same concerned diligence that he or she would a reported crime. The sifting of every shred of evidence to establish each element needed to prove the "*corpus delicti*" must be done in a systematic manner. By using the outlined material contained in this system, both the new and veteran officer should be able to successfully complete each investigation with a great deal less effort. He or she will provide a combination of systematic, uniform, and relevant, factual information needed by his or her department and that of the prosecuting agency.

DISCUSSION QUESTIONS

1. Outline the various procedures used in the investigation of a hit-and-run accident. What additional steps does the agency in your state require in completing such a detailed investigation?

2. Compare the uniform elements on establishing traffic accident violations with those of your state or its agency. How do they differ and what are the similarities if any?

True - False

1. Actual collision contact must be established in order to show that a hit-and-run accident actually occurred.

2. It is always a good idea to revisit the scene of a hit-and-run investigation at about the same time of day and on the same day of the week, in the possibility of locating the hit-and-run driver.

3. As a police officer, your opinions and conclusions in your accident report should never reflect how each element can be proven.

4. One method of interviewing a witness or driver is the question-answer technique.

5. If an officer does a complete job of investigation, it is seldom necessary to review the entire investigation prior to leaving the scene.

11: Determination of Speed from Skid Marks

Introduction to the Theory

In many investigations knowing the initial speed of the accident vehicle is of great value to the investigator. Drivers may not admit accurate speeds. Often the speed estimate of witness will be unreliable or disbelieved due to their unfamiliarity with speed and its results. A scientific method which eliminates much of the guesswork is that of computing minimum initial speeds from the measured length of skid marks.

Excessive speed is a common accident cause. When faced with an emergency, a driver will usually apply the brakes with maximum force or swerve the vehicle sharply. The resultant skid marks provide evidence which serves as a basis for computing the minimum initial speed of the vehicle with reasonable accuracy.

Not only will this evidence be of assistance in prosecuting violators, but it may also serve to exonerate drivers who might otherwise be charged. A skid of 20 feet from 20 miles per hour does not mean that a skid of 100 feet, or five times as long, indicates five times as much speed. The actual speed in this case would be only about 45 miles per hour.

This type of evidence is particuarly valuable in auto-pedestrian cases where little damage is done to the vehicle. Where extensive damage results, all that can be measured is the amount of speed the vehicle lost in sliding. Energy absorbed in the collision is above and beyond that figure.

With emergency stopping, rotation of the wheels ceases almost instantly and wheels are said to be "locked." Locking the wheels will not stop the vehicle at once. Instead, the tires will slide upon the surface until all the kinetic energy of the vehicle has been transformed into heat energy, represented by the sliding of the tires. Since, by the law of conservation of energy, energy cannot be created or destroyed, the energy of motion must be equal to the energy of stopping.

While this energy transformation is taking place, a heavy deposit of rubber will be left on the pavement. This can be readily identified, and the length of the skid mark left by each wheel can be measured.

Coefficient of Friction

Coefficient of friction is the ratio of the force required in sliding an object over a surface to the force (usually the weight) pressing the object against the surface.

The force created by locking the wheels of a moving vehicle is the same as that necessary to drag the vehicle with its wheels locked. The coefficient of friction (also called the drag factor) is found by using the formula of $f = F/W$ (also $F = Wf$). For example, if a force of 3,000 pounds were necessary to move a 4,000 pound vehicle with locked wheels, the coefficient of friction would be 75 percent.

$$f = \frac{F}{W} = \frac{3{,}000}{4{,}000} = 75 \text{ percent}$$

If the same vehicle were moving and the wheels were locked, a 3,000 pound force would "pull" on the vehicle until it stopped.

On dry, hard road surfaces braking forces which can be developed are fairly uniform. The coefficient of friction will vary slightly with changes in speed. In contrast, on wet surfaces, due to the lubricating effect of the water and other factors, the coefficient of friction will vary with different initial speeds. Therefore, use of the formulas for computing speed from skid marks should generally be confined to dry pavements.

Even with a freely rolling wheel, minute particles of rubber are being deposited on the surface of the pavement. However, these deposits are so small that no clearly defined markings are visible. With a heavy brake application, it is often difficult to detect any markings on the pavement if the wheels continue to rotate. Under emergency conditions, however, maximum braking force generally causes the wheels of the vehicle to lock. On dry pavement the locked wheels will leave a heavy deposit of rubber which can be accurately assessed as to length, location, direction, and number of wheels affected.

When a heavy braking force is exerted (even with disc brakes and/or radial tires), the retarding force is greatest just before the wheels are locked. This type of tire marking is termed "impending." The wheels are just about to cease rotating, and usually the tire marks are less visible than those made by locked wheels. In some instances it is possible to see the design of the tire tread in the impending markings. Usually this type of mark leads directly to the locked wheel markings. A general characteristic of the locked wheel type of markings is that the path of the vehicle is straight. This is true because it is impossible to change direction by steering once the wheels are locked.

In many collisions there may be a sudden and violent change of direction due to impact with another vehicle or object. This change is usually visible, but many times results in a confusion of tire markings which are difficult to describe accurately. In some instances, an abrupt change of direction is effected by the operator attempting to turn the vehicle from its path. It is possible when this occurs that clearly defined tire markings will be visible and that braking *may* or *may not* be in effect. In this case, a new force has been added to the vehicle which is described as "centrifugal force." This new force resists the change of direction which the operator is attempting to impart by turning the wheels.

Standard Tire Mark Terminology

Investigators should be aware that it is possible to leave tire markings on the pavement which closely resemble "locked wheel" or "impending" marks *without any brake application.* Such markings are many times identified by a variety of terms which are confusing. These types of descriptions do not lend themselves to intelligent interpretations or evaluations in reconstructing the facts of the collision.

In order to rectify this situation, a uniform terminology must be developed and used. To arrive at a conclusion as to speed based upon the evidence left on the roadway, it is necessary that all deceleration factors be considered. The primary deceleration factor is the braking mechanism. Secondary factors are the surface upon which the vehicle is traveling, its grade percentage, and its condition.

First among the uniform definitions of skid mark terminology is a *tire skid mark,* defined as a mark left by a tire due to the sliding of the tire over a surface. Visible tire skid marks are made in three ways: Extreme deceleration, extreme change of direction, and extreme acceleration.

Extreme Deceleration. When the braking system is sufficient to cause the wheels to cease rotating, they will lock and skid marks will be left on the surface. Even when the wheels are still turning but are *about to cease rotating,* skid marks will be visible.

Extreme Change of Direction. This action may result from effort on the part of the operator or an impact. The resulting skid marks are usually visible to the investigator.

Extreme Acceleration. This condition occurs when a propelling force exceeding the pavement efficiency is generated. If the driving force applied to the rear wheels is such that it will cause the wheels to "spin," skid marks will result.

The following diagrams illustrate some of the applications of the four basic terms which have been adopted and will be used throughout this discussion as defining the various types of tire skid marks.

Straight or curved skid marks due to brake application.

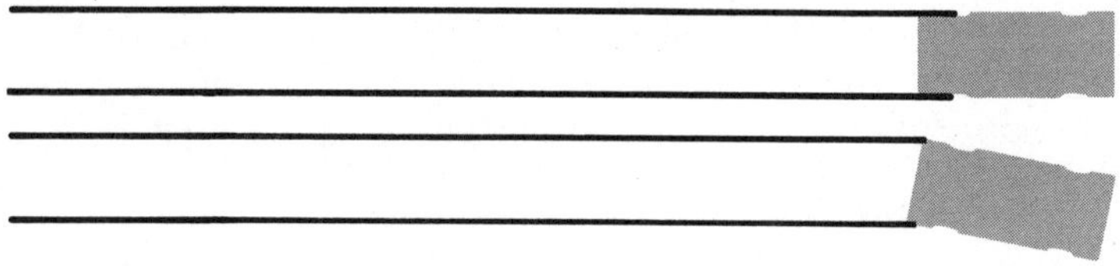

FIGURE 11-1

Impending skid mark: left by a braked wheel *rotating* slower than the forward motion of the vehicle in a straight or curved line.

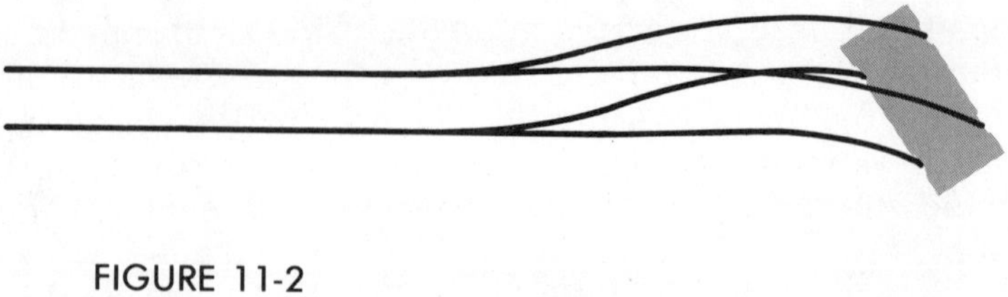

FIGURE 11-2

Locked wheel skid mark: left by a *nonrotating* wheel of a vehicle moving in a straight or curved line in the original direction of motion.

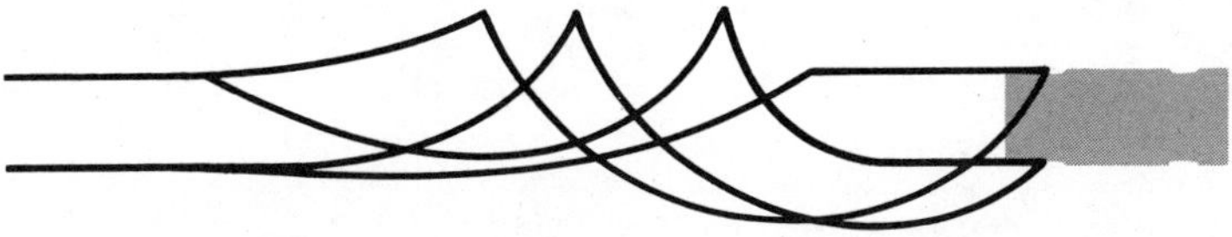

FIGURE 11-3

Slide skid mark: left by a *locked,* braked, or rotating wheel of a vehicle sliding in other than a forward direction, except when caused by centrifugal force.

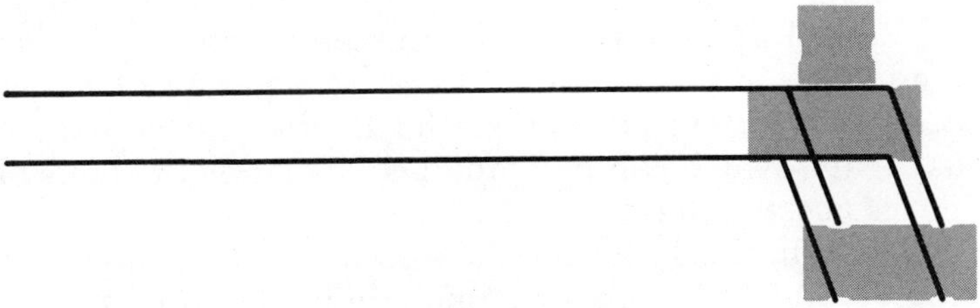

FIGURE 11-4

Side skid due to impact with brakes applied.

Centrifugal skid mark: left by a *rotating* wheel of a vehicle rounding a curve or turning at such a speed that centrifugal force entirely or partially overcomes the frictional resistance between road and tire.

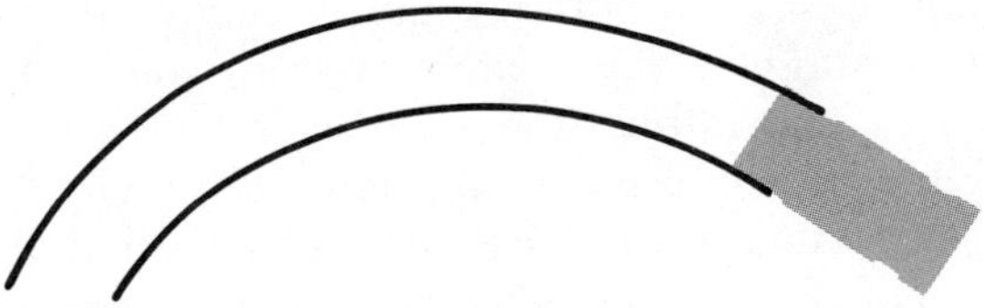

FIGURE 11-5

Curved light marks from a controlled rapid change of direction.

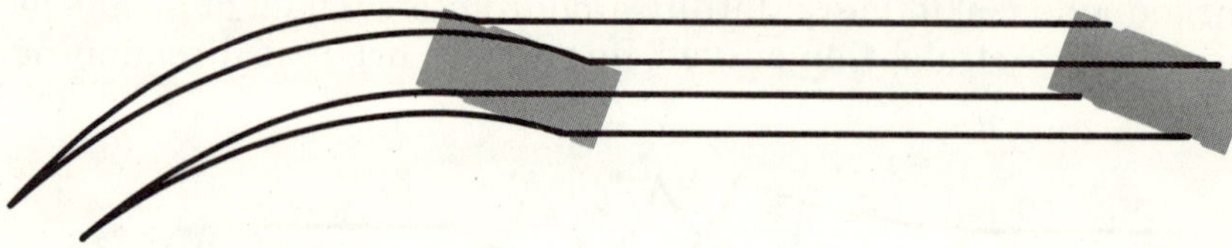

FIGURE 11-6

Curved marks from a too rapid change of direction. Skid mark ending with locked wheel or side mark.

There are descriptive terms which can be used to describe skid marks *after* the *basic term* has been used. These may be "intermittent" to describe markings which are not continuous; "dual wheel," "single vehicle," and so on to indicate the number of wheels affected and the type of equipment; "curving," "straight line," and so on to describe direction and type. Any term which further describes the skid marks after the basic terminology has been used, and which can be employed to clarify the facts of the collision, is acceptable.

Only when the basic terminology is used by all investigators will it be possible to reach a common understanding. Thus, definite conclusions may be made concerning speed, direction of travel, and point of impact from the skid marks at the accident scene.

Skid Mark Measurements

In order that such evidence may prove of value in determining minimum initial speeds and to assure its acceptance by the courts, it is imperative that it be properly obtained and recorded. The investigator is concerned with how far the vehicle skidded. Therefore, it is necessary to measure the mark left by each tire of the vehicle. In some cases this is fairly easy to do; in other instances it is almost impossible when the marks of front and rear wheels are superimposed. In the latter case, the investigator must measure the total length of the skid mark and deduct the wheel base of the vehicle. In any event, he or she must establish that the marks were left by a certain vehicle. This is usually accomplished by an admission from the driver, the testimony of a witness, or the wheels of the vehicle being atop each skid mark when at rest.

In measuring curved skids (other than centrifugal skids) the tape should follow the curvature of the skid mark, as the distance the wheels were in contact with the road surface is needed to calculate for speed. The investigator must see both ends of the tape in order for his or her testimony to be admissible.

Sometimes a locked wheel will bounce and leave a "skip skid"—a series of locked wheel skid marks with short spaces between them where the wheel was in the air. In these cases the distances during which the wheel was not in contact with the road surface should be included in the skid measurement. Although none of the speed is being dissipated while the wheel is in the air, the greater energy dissipated before the wheel leaves the road compensates for this. The skips will be only a few feet long.

In some instances a person will apply brakes when faced with an apparant danger, leaving skid marks, then release them as the situation changes, then apply them again—or the driver may "pump" the brakes. The skid marks will have a gap between them, but the gap will generally be more than ten feet in length. Each set of marks should be measured, added, and their total used for speed computation (the gaps are *not* included as is the case with skip skids).

Having measured the skid marks, the measurements must be properly recorded to be useful. For each wheel the length of the impending and locked wheel skids should be separately listed. It is also necessary to record the weather, the condition and the type of pavement, and the percentage of grade (noting whether it is up or down).

Calculating Speed from Skid Marks

When this information is secured and preserved it is simple to calculate the *minimum* initial speed of the vehicle. This can be done in several ways. What the investigator must remember is that *skid marks do not show all of the speed.* Neither the energy used in braking before the wheels began to leave visible marks, nor the energy left in the vehicle at the time of the collision will appear. What can be determined is the amount of speed lost in the distance the vehicle slid. The calculation will show less speed than the actual speed of the vehicle. Any deviation from the actual speed will, therefore, be in favor of the driver.

The simplest way to estimate the minimum initial speed of a vehicle traveling on a dry, level, hard-surfaced pavement is to arbitrarily assign a coefficient of friction of 60 percent. This is a safe minimum. The percentage of grade is added to this figure if uphill, or subtracted if downhill. Skids (impending *plus* locked wheel) from all skidding wheels should be averaged and the result used to calculate the vehicle's speed.

The formula used in calculating speed from skid marks is:

Basic Speed Skid Formula: $$S = \frac{V^2}{30f}$$

"S" represents the skid distance in feet, "V" (velocity) the speed in miles per hour, and the "f" the coefficient of friction. This formula may be transposed to read $V^2 = 30Sf$ or $V = \sqrt{30Sf}$, which are easier to use. What the investigator is concerned with is the speed. He or she simply substitutes known factors in the right-hand side of the equation, performs the multiplications, extracts the square root, and obtains the minimum speed. This can be easily done in a matter of a few seconds at the accident scene which gives the investigator a result close enough to guide his or her investigation. It is a quick method of determining whether there was sufficient speed involved to warrant running test skids or further analysis.

If there is indication of excessive speed, particularly in a serious accident, test skids should be made to assure that the evidence will be as good as possible in the event of prosecution. Although it is true that the weight of the vehicle involved, the condition of tires, or the condition of brakes (provided they are sufficient to lock the wheels) is not significant, the task of convincing a jury is much simpler if test skids are made by the same vehicle which was involved in the accident. In some cases damage may prevent its use; however, if possible the vehicle involved should be used. The test should be made at the same location and in the same direction the accident vehicle was traveling. A minimum of three test skids from the same speed should be used in running test skids due to the danger involved. For all practical purposes, thirty miles per hour should be fast enough.

As an example of the use of skid marks in determining initial speeds, the following will illustrate several principles. The investigator has measured 50 feet of accident skid marks, locked wheel plus impending, on a dry, level asphaltic concrete pavement. The driver involved states he was going 20 to 25 miles per hour. The quick method of estimating initial minimum speed is $V = 30 \times 50 \times .60$, or 900. The square root of 900 is 30. Therefore, the minimum speed is 30 miles per hour. This can be done simply and without a test skid by arbitrarily assigning a 60 percent coefficient of friction.

Desiring to arrive at a true coefficient of friction, the investigator will run three test skids from the admitted speed of 25 miles per hour. These tests give skid distances of 28, 29, and 30 feet respectively. The longest skid from the claimed speed is 20 feet shorter than the accident skid (see fig. 11-7). This should be quite convincing to the driver that he was traveling faster than he ad-

The investigator notes the necessary information, records his or her test skids, and arrives directly at the minimum speed of the accident vehicle by the following method:

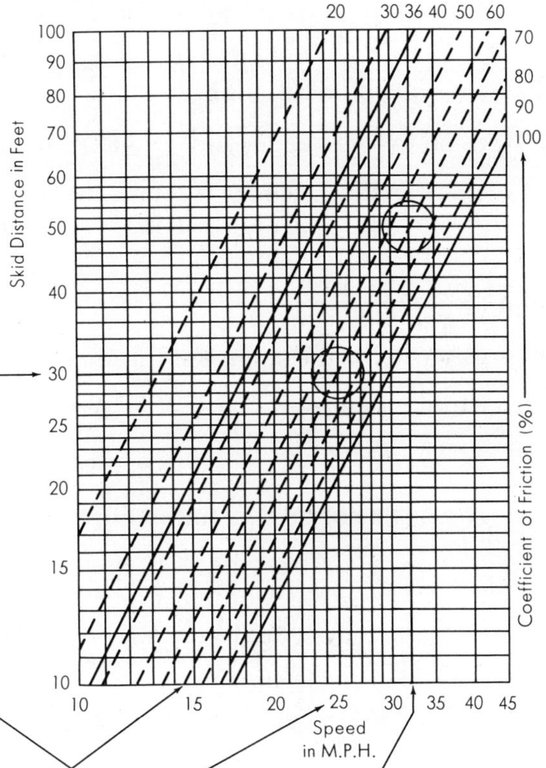

Step 1
On the "Skid Distance in Feet" axis he or she plots the longest test skid, in the above instance 30 feet.

Step 2
On the "Speed in M.P.H." axis of the chart he or she runs a line up at 25 miles per hour, which was the test speed.

Step 3
These lines intersect at the broken line indicating 70 percent coefficient of friction.

Step 4
The investigator then, using the same steps as indicated above, plots the 50 feet of accident skid on the "Skid Distance in Feet" axis. From the point where this line intersects the 70 percent "Coefficient of Friction" line, he or she draws a line to the "Speed in M.P.H." axis and reads a minimum speed of 32 miles per hour.

FIGURE 11-7
Calculating Speed from Skid Marks

mits, particularly if the investigator points out the discrepancy in the skidding distance.

Formulas on Speed and Stopping of Motor Vehicles

The following chart of equations will show how the speed skid formulas are obtained.

Feet per Second and Miles per Hour

(1) Speed in feet per second = 1.467 times speed in miles per hour

60 mph × 1.467 = 88 ft./sec.
40 mph × 1.467 = 58.7 ft./sec.
20 mph × 1.467 = 29.4 ft./sec.

Energy and Speed

Energy varies as the square of the speed

(2) Kinetic Energy $= \frac{WV^2}{30}$ W = Weight of car in pounds
V = Speed in mph
Kinetic energy is in foot pounds

A 3000 lb. car has the following energies:

At 60 mph, K.E. = 360,000 ft. lbs.
40 mph, K.E. = 160,000 ft. lbs.
20 mph, K.E. = 40,000 ft. lbs.

Speed and Falling Distances

(3) $V^2 = 30\ S$ S = Height of fall in feet
V = Speed in mph

When Speed V = 60 mph, S = 120 ft.
V = 40 mph, S = 53.3 ft.
V = 20 mph, S = 13.3 ft.

Example: Running into a solid brick wall at 60 mph will do as much damage as running off a building 120 feet high.

Speed and Stopping Distances

Total stopping distance equals braking distance plus reaction distance

(4) Minimum original speed = Speed of car in test

$$X \sqrt{\frac{\text{Length of original skidmarks}}{\text{Braking distances in test}}}$$

Example: Accident skidmarks average 60.0 feet.
Test skid from 30 mph gives braking distance of 40.0 feet.

$$\text{Then Speed} = 30 \sqrt{\frac{60}{40}} = 36.8 \text{ mph minimum}$$

Car must have been going at least this fast before the accident. It probably was going even faster.

This formula also can be written as:

$$V \text{ (original)} = V \text{ (of test)} \quad X \sqrt{\frac{s \text{ (measured)}}{s \text{ (test)}}}$$

s (measured) = average length of four skid marks following the accident.
s (test) = braking distance as measured by pistol, for test run.
V (test) = speed of test run in mph
V (original) = computed minimum speed of the accident vehicle (the vehicle probably was going quite a bit faster).

Speed, Friction and Braking Distance

(5) $V = 5.5 X \sqrt{fs}$

(6) $f = \dfrac{V^2}{30s}$

(7) $s = \dfrac{V^2}{30f}$

V = Speed on mph
s = braking distance in feet.
If wheels skid, f = coefficient of friction
If wheels do not skid,

$$f = \frac{\text{Percent of braking}}{100}$$

Speed on Curves

(8) $V = 15\, r\, (e + f)$

V = Maximum speed in mph
r = radius of curve in feet
e = superelevation in feet per foot
f = sideways friction coefficient

For most highway curves, e is between zero and 0.10. For comfort in rounding curves, f should not exceed 0.20.

A car will skid sideways when f approaches the coefficient of friction of the pavement

A method which saves one operation is by direct substitution in the formula $V_a = V_t (S_a/S_t)$, where V_a equals accident speed, V_t equals test speed, S_a equals accident skid distance, and S_t equals test skid distance. This, too, gives a result of 32 miles per hour. It can be seen by the above that the quick approximation with the assumed coefficient of friction of 60 percent gives a result close enough for practical use.

In using a test skid, it is important that every effort be made to get all the impending skid; otherwise, the value of the coefficient of friction obtained will be too high. It is necessary that the speedometer be accurate so that the evidence obtained will be reliable for presentation in court.

While the investigator will be able to calculate initial speeds closely by these methods, presentation in court will probably require expert testimony, including explanation of the various physical laws and mathematical formulas involved. The expert will rely on the information given him or her by the investigator to reach his or her conclusion. This information must be accurate, understandable, and complete, for without this information skid mark evidence will be of little value. With the proper measuring and recording of skid mark evidence, accurate minimum initial speeds may be computed.

In complicated cases where not all the wheels slide, or where skid marks are on different types of surfaces, it will probably be up to the expert to determine if any use can be made of the skid mark evidence. However, it is important to record the evidence so that the expert may decide to what extent it is usable.

When a vehicle is rounding a curve, there are two opposed forces acting upon it. Centrifugal force acts to keep the vehicle moving in a straight line, while centripetal force is being applied to turn the vehicle. If the centrifugal force is greater than the frictional resistance between the tires and the pavement, the vehicle will skid. Skid marks left by a vehicle sliding on a curve can be used to tell something about how fast a vehicle was going, for centrifugal force is directly proportional to the weight of a body and the square of the velocity and is inversely proportional to the radius of the curvature of its path. The speed at which a curve can be rounded without skidding is known as the *critical speed.* When a vehicle skids on a curve without brake application it is because the critical speed was exceeded. *By calculating the critical speed of a curve the investigator will know the minimum speed at which the vehicle must have been traveling in order to slide.*

To calculate the critical speed of a curve, the investigator must know the *radius* of the curve, the bank or superelevation, and the *coefficient of friction.*

The radius of a curve is determined as follows:

1. Mark two points on the curve using the outside edge, as the inside of curves are frequently widened to make them safer. The points used can be where the curve straightens or two points 100 feet or less apart near the accident. Use a long distance if the curve is a big one.

2. Measure the distance between these marks, but instead of rolling up the tape, release both ends of the tape carefully without moving it. Let it lie flat and straight on the ground. The distance so measured is called the chord. It is "C" in the formula below.

3. Divide the length of the chord by two, and without moving the tape, find the middle of the chord. Mark the spot.

4. Carefully measure the shortest distance from the middle of the chord to the curve. Use a pocket tape if available. Measure to the nearest half inch. This distance is called the middle ordinate. It is "h" in the following formula.

Example: Measure straight across part of a circular curve with a chord of 50 feet, "C." The distance from the middle of the chord (at 25 feet) to the curve is one foot. This is the middle ordinate, "h." Using these figures in either of the two following formulas, we are able to determine the radius.

$$\text{Radius} = \frac{C^2}{8h} + \frac{h}{2} \quad \text{(in this formula the middle ordinate is in feet)}$$

$$\text{Radius} = \frac{50 \times 50}{8 \times 1} + \frac{1}{2} = \frac{2500}{8} + \frac{1}{2} = 312.5 + .05 = 313 \text{ feet}$$

OR

$$\text{Radius} = \frac{30^2}{2h} \times \frac{h}{24} \quad \text{(in this formula the middle ordinate is in inches)}$$

$$\text{Radius} = \frac{3 \times 2500}{2 \times 12} + \frac{12}{24} = 7500 + \frac{1}{2} = 312.5 + 0.5 = 313 \text{ feet}$$

It may prove easier to use a calculator which works the formula for the investigator in a quicker manner (fig. 11-8). By drawing a straight line across the chart we can find the radius of the circle of which the measured curve is a part. When we use two known points (length of chord and middle ordinate), we simply connect the two on the radius calculator and read the answer on the right column (radius, feet). Using the calculator and a set of known measurements, find the radius in feet.

Example: Use the same measurements as in the previous example; chord 50 feet and middle ordinate 1 foot. Find the chord, 50, on the left scale, C. Find the middle ordinate, one foot, on the middle scale, h. Draw a *straight* line from the 50 on scale C through 1 on scale h to scale R. Read the radius on scale R at this point. It is 313 feet.

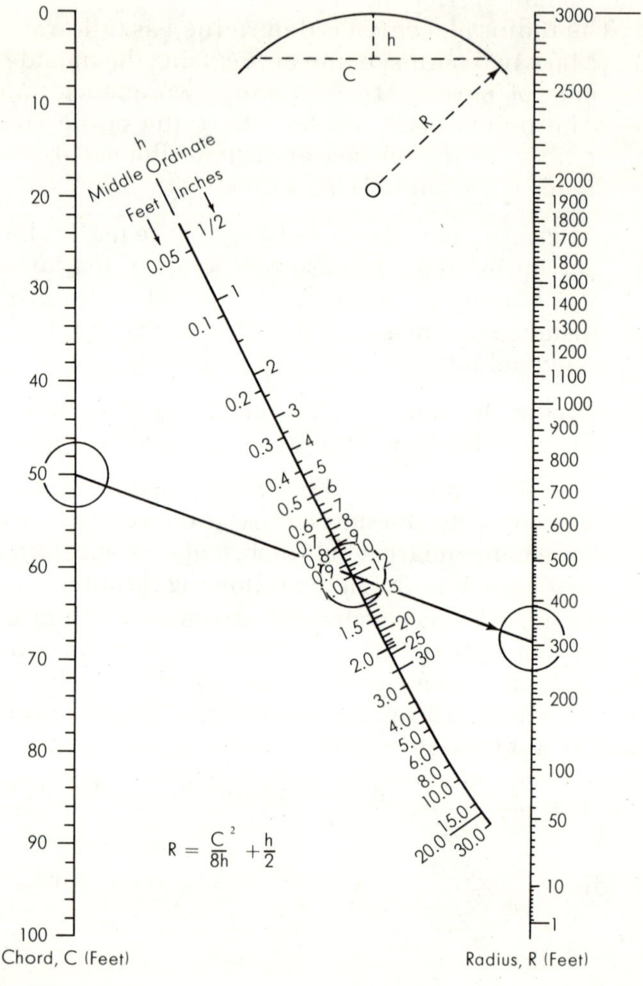

FIGURE 11-8
Radius Calculator

This same formula and calculator can also be used to find the radius of a small circle by using inches instead of feet for measurements. This might help the investigator find the size of a round headlight lens from a piece of glass found at the scene of an accident.

Coefficient of friction may be determined as previously described. Superlevation is added or subtracted just as grade percentage is in other skid mark problems. (If test skids are made, the superelevation or grade is already included in the resultant coefficient of friction.)

Superelevation or grade percentage means the steepness of the road banking or a hill. They are both measured by the number of feet the road rises for each foot of level distance along or across the road. The resulting number is less than 1.00, and, except on steep hills or banks, is less than 0.10.

When drawing a diagram of an accident scene, always draw an arrow pointing *uphill* after the grade number. Also, since uphill is plus grade, put the plus sign (+) in front of the number. *Downhill* or negative grades are written with a *minus* sign (–) in front of the number. Whether a grade is positive or negative depends upon the direction of travel (see fig. 11-9).

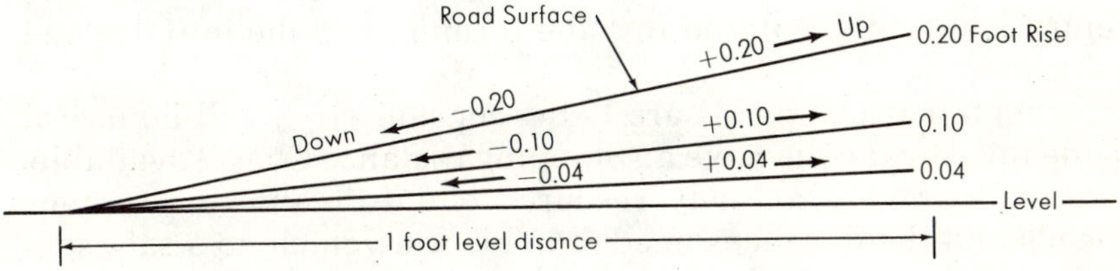

FIGURE 11-9
Superelevation of Road

The grade or slope of a road is the rise divided by the level distance. The bank or superelevation is the same thing *across* the road. The number representing grade is often mentioned in terms of feet of rise or fall per hundred feet of horizontal distance. It then becomes feet per hundred or grade, or percent grade.

Example: A road rises two feet in a level distance of 50 feet. Divide 2 by 50. The result is 0.04. (This is 4 feet per hundred, or a 4 percent grade). This information would be indicated on a diagram by an arrow pointing *uphill* and writing +0.04 grade. This percentage of grade would be added to the coefficient of friction when computing speed from skid marks.

The formula to calculate critical speed is $V^2 = 15 \times R \times f$, or $V = \sqrt{15Rf}$, where V equals speed, R equals radius, and f is the coefficient of friction.

Example: Calculate critical speed where the chord measures 50 feet, the middle ordinate is 11 inches, the coefficient of friction is 68 percent, and the superelevation is 2 percent.

$$R = \frac{3 \times 2500}{22} + \frac{11}{24} = 341 \text{ feet}$$

$$f = .68 + .02 = .70$$

$$V^2 = 15 \times 341 \times 170$$

$$V = \sqrt{3580.50}$$

$$F = 59.8 \text{ mph}$$

Where brakes are applied during a centrifugal skid, the formula will not hold true, however, as brake application will cause skidding at lower speeds than if brakes were not applied.

The same formula used to calculate the critical speed of a curve is used to compute minimum initial speed from centrifugal skid marks. The centrifugal skid mark is the curve on which the chord and middle ordinate are measured. The radius of the curve of the centrifugal skid is entered into the formula and the initial speed calculated.

No matter what speeds are being considered, it will be useful to the investigation to have a "Stopping Distance Chart" available. This will provide a ready resource of information on various speeds and the distances needed to bring a vehicle to a safe stop upon perceiving a roadway hazard (see fig. 7-14).

Summary

The investigator who uses all the tools available to him or her will certainly give expert credence to his or her investigation and to his or her eventual testimony in court. Part of those tools are the various formulas used in determining speed from skids. When properly applied, the investigator becomes a valuable key in the investigation of many complicated accidents. The use of testimony in describing skid marks and speed estimates may be admissible providing the proper identifying foundation is laid. Although the individual states have varying evidence codes, most all of them subscribe to the prevailing rule asserting that the weight of the testimony by an investigator in establishing his or her expertise is based on the individual court's legal parameters setting forth the acceptable background for such technical testimony, and the jury's ability to assess that testimony in the end. The individual officer who applies these rules to his or her investigations will find a deep sense of reward in knowing that he or she is filling a critical need in the field of accident reconstruction. The expert testifying as to speeds determined from skid marks should know that it is reversible error to withhold such testimony from the court when it is properly offered in evidence. True, the validity of the speed estimate is dependent on a number of important factors, such as the drag factor of the particular highway in question upon which the skidding took place, the accuracy of the skilled investigator to reproduce the accident conditions as exactly as possible during test skids, and, finally, his or her mathematical ability to compute the drag factor and the estimate of minimum speed.[1]

To the new officer just learning accident investigation theory and faced with the challenge of developing the necessary skills in determining speed from skids, this may seem an almost endless task. Yet, with perseverance and time one may become a department's leading expert in the field of speed-skid analysis. To the seasoned veteran who has worked hundreds of traffic accidents, but who has never had to testify as to the estimated speed of a vehicle prior to impact, he or she likewise should prepare for the possibility of having to give such expert testimony, for it should be his or her professional concern to do so should the occasion arise. Not only is it important for the individual officer to develop the necessary vocabulary of the terms and formulas used in speed-skid analysis, as well as a workable understanding of their physical application, but it is also important to continually use these scientific tools to assist in effectively bringing to justice those persons who have violated the law in the operation of their motor vehicle and which when added to the speed factor resulted in a traffic accident causing loss of life and/or property.

The professional traffic accident investigator will continually strive to see that the integrity of his or her testimony is always above reproach and that he or she has constantly prepared himself or herself to supply the factual evidence needed in assisting with the prosecution of his or her department's traffic accident cases.

---

1. James P. Hall, *Skid Marks: To Estimate Speed; Police Work* (Milwaukee, Wis.: Milwaukee Area Technical College, 1971).

DISCUSSION QUESTIONS

1. Discuss the theory of determining minimum speed from skids, injecting properly the various technical terms applied to speed-skid investigation and analysis.

2. Discuss the formulas for determining speed from skids, by identifying the methodology by which they are derived.

True - False

1. The term "Coefficient of Friction" is synonymous with drag factor.

2. It is not always necessary to identify and measure all of the impending skid left by the vehicle in question in order to obtain an accurate answer as to minimal speed.

3. The term "Kinetic Energy" can best be described as energy a body or mass has because of its motion.

4. Centripetal and Centrifugal forces are applied to a vehicle's ability to negotiate a curve.

5. A Radius Calculator is used in determining a roadway curve's "Critical Speed."

6. Superelevation is a term applied to determining the slope of a grade.

Bibliography

Adams, Thomas F. *Police Patrol Tactics and Techniques*. Englewood Cliffs, N.J.: Prentice-Hall, 1971.

American Association of State Highway Officials. *Manual on Uniform Traffic Control Devices for Streets and Highways*. Evanston, Ill.: Institute of Traffic Engineers, National Conference on Street and Highway Safety, 1969.

Baker, J. Stannard. *Traffic Accident Investigator's Manual For Police*. Evanston, Ill.: Traffic Institute, Northwestern University, 1966.

Baker, J. Stannard, and Stebbins, William R., Jr. *Dictionary of Highway Traffic*. Evanston, Ill.: Traffic Institute, Northwestern University, 1964.

California Highway Patrol. *Accident Investigation*. Sacramento, 1970.

Committee on Highway Capacity of the Highway Research Board. *Highway Capacity Manual*. National Research Council, U.S. Bureau of Public Roads. Washington D.C.: Government Printing Office, 1960.

Daily Training Bulletin of the Los Angeles Police Department. Bulletins 1-173. Springfield, Ill.: Charles C. Thomas, 1963.

Department of California Highway Patrol. *Traffic Accident Investigation Manual*. Sacramento, 1966.

Department of Highway Transportation, Division of Highways, State of California. *The Highway Engineer*. Sacramento, Calif.: State Printing Office, 1966.

Donigan, Robert L., and Fisher, Edward C. *The Evidence Handbook*. Evanston, Ill.: Traffic Institute, Northwestern University, 1965.

Horgan, John J. *Criminal Investigation*. New York: McGraw-Hill, 1974.

International Association of Chiefs of Police. *Municipal Traffic Supervision Manual*. Evanston, Ill.: Traffic Institute, Northwestern University, 1971.

International City Managers Association. *Municipal Police Administration*. Chicago, Ill.: Institute for Training in Municipal Administration, 1969.

National Committee on Uniform Traffic Laws and Ordinances. *Uniform Vehicle Code and Model Traffic Ordinance*. Washington, D.C., 1972.

O'Brien, Kevin P., and Sullivan, Robert C. *Criminalistics—Theory and Practice*. Boston, Mass.: Holbrook Press, 1972.

Schwartz, Louis B., and Goldstein, Stephen R. *Law Enforcement Handbook for Police*. St. Paul, Minn.: West Publishing, 1970.

Soderman, Harry, and O'Connell, John H. *Modern Criminal Investigation*. New York: Funk and Wagnalls, 1962.

Stuckey, Gilbert B. *Evidence for the Law Enforcement Officer*. New York: McGraw-Hill, 1974.

Traffic Accident Investigators Manual for Police. Evanston, Ill.: Traffic Institute, Northwestern University, 1963.

U.S., Department of Health, Education, and Welfare. *Uniform Definitions of Motor Vehicle Accidents*. Washington, D.C.: Government Printing Office, 1971.

Vanderbosch, Charles G. *Criminal Investigation*. Washington, D.C.: International Association of Chiefs of Police, 1968.

Weston, Paul R., and Wells, Kenneth M. *Criminal Evidence for Police*. Englewood Cliffs, N.J.: Prentice-Hall, 1971.

———. *Elements of Criminal Investigation*. Englewood Cliffs, N.J.: Prentice-Hall, 1971.

Whisenand, Paul M., and Cline, James L. *Patrol Operations*. Englewood Cliffs, N.J.: Prentice-Hall, 1971.

Determining Speed from Skid Marks

Charts and Tables for Stopping Distances of Motor Vehicles. Evanston, Ill.: Traffic Institute, Northwestern University, 1960 and 1966.

Department of California Highway Patrol. *Speed From Skidmarks*. Sacramento, 1968.

Dull, Charles E. *Modern Physics*. New York: Henry Holt, 1945.

Freeman, Ira M. *Modern Introductory Physics*. 2d ed. New York: McGraw-Hill, 1957.

Rogers, Eric M. *Physics for the Inquiring Mind.* Princeton: Princeton University Press, 1960.

Srack, C. Leonard. *Miles-Feet-Seconds.* Los Angeles, 1951.

Uses of Skidmarks in Traffic Cases. Evanston, Ill.: Traffic Institute, Northwestern University, 1969.

Traffic Records Systems

Probably the most comprehensive and analytical traffic records system in the world is the CARES/ACTRS system utilized by a number of communities in the San Francisco, California, Bay Area. The system was developed by Stanford Research Institute under contract to the cognizant communities. The system has been thoroughly documented, and anyone seriously contemplating the development of a traffic records system is advised to review this system. All the documents can be obtained from SRI in Menlo Park, California.

Cook, D.W., and Schlaefli, J.L. *ACTRS Citation System Users Manual.* SRI Technical Note 1011-1, May 1972.

Marcum, K.R., and Schlaefli, J.L. *ACTRS Sign System Users Manual.* SRI Technical Note 1261-1, February 1972.

Sandys, R.C., and Schlaefli, J.L. *ACTRS: Alameda County Traffic Reporting and Inventory System Procedures Manual.* SRI Project 1542, May 1972.

Schlaefli, J.L., and Bagley, M.D. *San Mateo County Computerized Accident Records System (CARES).* Final Report of SRI Project 1077, June 1972.

Schlaefli, J.L. et al. *San Mateo County Computerized Accident Records System (CARES) Systems Manuals—Volume I: File Descriptions and Maintenance Procedures.* SRI Technical Note 1077-2, June 1972.

———. *Alameda Countywide Traffic Records System (ACTRS) User's Manual.* SRI Technical Note 8513-1, October 1971.

———. *San Mateo County Computerized Accident Records System (CARES) User's Manual.* SRI Technical Note 1077-1, June 1972.

For a detailed description of the traffic records system implemented in the Los Angeles Police Department in the early 1970s, containing numerous flow charts and sample print-outs, see:

Optner, Stanford L. & Associates, Inc. *Traffic Information System (Phase III).* Los Angeles, 15 May 1970.

Other books treating traffic records systems:

Eastman, George D., and Eastman, Esther M., eds. *Municipal Police Administration.* Washington, D.C.: International City Management Association, 1969. This basic treatise of police administration discusses traffic records in chapters 6 and 14.

Hewitt, William H. *Police Records Administration.* Rochester, N.Y.: The Lawyers Cooperative Publishing Company, 1968. Chapter 16 discusses in detail the legal aspects of traffic records and the essentials for documenting accidents and citations.

Whisenand, Paul M., and Tamaru, Tug T. *Automated Police Information Systems.* New York: John Wiley, 1970. A systems view of police information systems as an integral part of the overall criminal justice system. While a very theoretical treatment of police information systems, this book contains an extensive bibliography of literature about automated police information systems in existence in 1968.

Books about electronic data processing (EDP) and computers:

Benice, Daniel D. *Introduction to Computers and Data Processing.* Englewood Cliffs, N.J.: Prentice-Hall, 1970. This introduction to EDP emphasizes the programming aspect and introduces the fundamentals of several current programming languages: System/360 Assembler, FORTRAN, COBOL, PL/1, and RPG.

Berkewitz, Nathan, and Munro, Robertson, Jr. *Automatic Data Processing and Management.* Belmont, Calif.: Dickenson, 1969. An introduction to EDP that emphasizes the implications for managers and what their responsibilities should be.

Harvard Business Review. *Computer Management Series: Reprints from Harvard Business Review.* A collection of sixteen articles appearing in HBR from 1962 through 1967. While these articles are slanted toward business uses of computers, the information can be invaluable to the officer involved in the administration of an automated traffic records system.

Martin, E. Wainwright, Jr. *Electronic Data Processing: An Introduction.* Homewood, Ill.: Richard D. Irwin, 1965. A textbook for an introductory course in EDP, the book was designed for the mature student (graduate or advanced undergraduate level). It contains the essentials for the officer who is responsible for EDP application in his or her department.

Odell, Margaret K., and Strong, Earl P. *Records Management and Filing Operations.* New York: McGraw-Hill, 1947. A thorough treatment of manual record keeping which serves as a guide to records department operations. It is especially useful for improving the computerized records functions that exist alongside a computerized system, or for the small police department that cannot afford a computer.

Index